INTRODUCTION TO
GROUP THEORY

INTRODUCTION TO
GROUP THEORY

W. LEDERMANN

OLIVER & BOYD · EDINBURGH

OLIVER & BOYD
Croythorn House
23 Ravelston Terrace
Edinburgh EH4 3TJ
A Division of Longman Group Limited

ISBN 0 05 002652 6

Printed in Great Britain by
Bell & Bain Ltd. Glasgow

Contents

V. Generators and Relations

VI. Series of Subgroups

VII. Permutation Groups

VIII. Sylow's Theorems

Preface

In the twenty-five years since my *Introduction to the Theory of Finite Groups* was first published, the teaching of Group Theory has become greatly extended and diversified: it is now studied by all mathematical undergraduates, and the fundamental concepts of the subject are part of the teacher's training at Colleges of Education, whilst in a modern school syllabus the theory of groups usually turns out to be one of the more popular items. In view of this lively and widespread interest in groups it is not surprising that the text has shown signs of ageing, which could not easily be rectified by a revised edition.

A fresh start has therefore been made with the present *Introduction to Group Theory*: the nomenclature and notations have been brought up to date, less emphasis is laid on finite groups (as indicated in the title) and a number of additional topics have been included, albeit briefly, such as central series and nilpotent groups. Despite these changes I have tried to preserve the elementary character of the older book. The earlier chapters should be accessible to a sixth form pupil with an enquiring mind, and the entire book is intended to cover the bulk of the work on group theory in an Honours course. As before, I have not always chosen the shortest path to reach a particular goal, when I believed that an alternative route was more instructive and rewarding. On p. 173 I mention some more advanced and substantial text-books which, I hope, the reader will consult for a more profound study of group theory.

Throughout the years I have received numerous suggestions and criticisms regarding the former book. All these remarks have been useful and, whenever possible, have been incorporated in this volume. My especial thanks, however, are due to Professor J. A. Green. He has read the typescript with great care and has sent me invaluable comments which reflect his outstanding expertise and experience in this field.

Finally, I should like to thank the publishers for their courtesy and co-operation.

WALTER LEDERMANN

I. The Group Concepts

1. Introduction. The elementary operations of arithmetic consist in combining two numbers a and b in accordance with some well-defined rules so as to produce a unique number c. For instance, if the **law of composition** is multiplication, we should have that $c = ab$. When a and b are given, the number c can be found in each case.

It is known that multiplication of two or more numbers obeys certain formal laws which hold for all products, irrespective of their numerical values, thus:

$$ab = ba; \quad \textbf{(commutative law)} \quad (1.1)$$

$$(ab)c = a(bc); \quad \textbf{(associative law)} \quad (1.2)$$

$$1a = a1 = a. \quad (1.3)$$

The last equation serves to introduce one particular number called unity. The second law states more explicitly that, if we put $ab = s$ and $bc = t$, then it is always true that $sc = at$.

In the axiomatic treatment of arithmetic it is customary to begin by laying down postulates or axioms such as (1.1), (1.2) and (1.3), and certain others dealing with addition as well as multiplication, and then to deduce the logical consequences of these postulates. It is immaterial, at the outset, whether the symbols $a, b, \ldots$ represent numbers as we normally understand them or other mathematical entities or indeed whether they admit of any concrete interpretation. Numerous axiomatic systems are logically possible, but they are not equally interesting or significant. It is the variety and depth of applications in pure and applied mathematics that has caused one conceivable system of axioms to be preferred to another.

2. The Axioms of Group Theory. The abstract theory of groups deals with a finite or infinite set of elements

$$G: a, b, c, \ldots$$

with respect to which a single law of composition is defined. It is a matter of convention that the notation and nomenclature of multiplication is usually, though not always, adopted to express the composition of elements. Thus we assume that any two elements a, b of G, equal or unequal, possess a unique product c, and we write

$$ab = c.$$

In more formal language it is stated that with every ordered pair (a, b) of elements there is associated a unique element c, the term ordered pair meaning that, when $a \neq b$, we must distinguish between the pairs (a, b) and (b, a). It is an essential feature of a group that the product of two elements is again an element of the group, or in more technical parlance, that the group is closed with respect to multiplication. The type of multiplication used in groups must obey certain axioms, which are set out in the following definition.

DEFINITION 1: *A set G for which a law of composition ('multiplication') is defined, forms a group if the following conditions are satisfied:*

I. **Closure:** *to every ordered pair a, b of G there belongs a unique element c of G, written*

$$c = ab,$$

which is called the product of a and b.

II. **Associative law:** *if a, b, c are any three elements of G, then*

$$(ab)c = a(bc),$$

so that either side may be denoted by abc.

III. **Unit element:** *The set G contains an element 1, called* **unit element** *(or* **identity** *or* **neutral element***) such that for every element a of G*

$$a1 = 1a = a.$$

IV. **Inverse element:** *corresponding to every element a of G, there exists in G an element a^{-1} such that*

$$aa^{-1} = a^{-1}a = 1.$$

It will be observed that these postulates closely resemble those which govern multiplication in familiar number systems, for example the rational numbers, except that the commutative law is not in general required to hold for groups.

DEFINITION 2: *A group which has the additional property that for every two of its elements*

$$ab = ba$$

is called an **Abelian*** (*or* **commutative**) *group*.

The waiving of the commutative law for groups makes it necessary to distinguish between ab and ba, and we say that a has been post-multiplied or premultiplied by b respectively. Whilst the commutative law need not hold throughout the group, it may nevertheless be valid for some particular pairs of elements.

DEFINITION 3. *The elements a, b are said to commute (or to be permutable) if*

$$ab = ba.$$

For example, 1 commutes with every element, and a always commutes with a^{-1}, as demanded in IV.

We shall now draw some conclusions from the axioms which will shed further light on the group structure.

(i) The associative law was postulated only for three elements. But it will be seen that a product of n factors (given in a definite order) has a unique meaning, so that brackets may be inserted or omitted at will as long as the factors remain in the given order. For, using axiom II as basis of induction, we may assume that a product of fewer than n factors has already been defined, and that

$$a_1 a_2 \ldots a_r = (a_1 a_2 \ldots a_s)(a_{s+1} \ldots a_r),$$

where $1 < s < r < n$. It is required to show that

$$(a_1 \ldots a_r)(a_{r+1} \ldots a_n) = (a_1 \ldots a_s)(a_{s+1} \ldots a_n), \qquad (1.4)$$

which will imply that any two modes of brackets lead to the same result. The left-hand side of (1.4) can be written as

$$[(a_1 \ldots a_s)(a_{s+1} \ldots a_r)](a_{r+1} \ldots a_n) = [b_1 b_2] b_3,$$

where the products in round brackets are denoted by b_1, b_2 and b_3 respectively. The right-hand side of (1.4) can be expressed as

$$(a_1 \ldots a_s)[(a_{s+1} \ldots a_r)(a_{r+1} \ldots a_n)] = b_1 [b_2 b_3],$$

after the second factor has been broken up by virtue of the inductive hypothesis. By axiom II we have that

$$[b_1 b_2] b_3 = b_1 [b_2 b_3],$$

* After N. H. Abel (1802–29).

which proves the assertion (1.4). We are therefore entitled to omit the brackets altogether and denote either side by

$$a_1 a_2 \ldots a_n.$$

In particular, when all factors are identical we shall, as in ordinary algebra, write

$$aa \qquad\qquad = a^2,$$

$$(aa)a = a(aa) = a^3$$

....

Hence, when n and m are positive integers, we have that

$$a^m a^n = a^n a^m = a^{m+n} \qquad (1.5)$$

and

$$(a^m)^n = a^{mn}. \qquad (1.6)$$

It is interesting to observe that the familiar laws of indices (1.5) and (1.6) ultimately rest on the associative law of multiplication.

However, when a and b do not commute, it will in general be found that

$$(ab)^n \neq a^n b^n.$$

But, when a and b do commute,

$$(ab)^n = abab \ldots ab = a^n b^n \qquad (1.7)$$

and

$$a^m b^n = b^n a^m,$$

since in this case we may arrange the factors as we please.

(ii) Axiom III postulates the existence of a two-sided unit element. We shall now prove that there can be only one such element. For suppose that $1'$ is another element, having the same properties as 1. Then $11' = 1$, because $1'$ acts as a right unit on 1 and $11' = 1'$, because 1 acts as a left unit on $1'$. Thus $1 = 1'$.

(iii) The (two-sided) inverse postulated in Axiom IV is unique. For suppose that $aa_1 = 1$. Then $a^{-1}aa_1$ can be evaluated in two ways, namely

$$a^{-1}aa_1 = (a^{-1}a)a_1 = 1a_1 = a_1$$

and

$$a^{-1}aa_1 = a^{-1}(aa_1) = a^{-1}1 = a^{-1},$$

whence $a_1 = a^{-1}$. Similarly, the equation $a_2 a = 1$ implies that $a_2 = a^{-1}$. In fact, we have proved that any left inverse and any right inverse of a is equal to a^{-1}.

The equations
$$ax = b, \quad ya = b$$
have solutions
$$x = a^{-1}b, \quad y = ba^{-1}$$
respectively. In general $x \neq y$, and we have to distinguish between left and right 'division' by a. These solutions are unique, for if
$$ax = ax_1 = b,$$
left multiplication by a^{-1} yields $x = x_1$. Similarly, if
$$ya = y_1a = b$$
we infer that $y = y_1$.

Put in a different way, we may state that in every group the **cancellation law** holds, with regard both to left cancellation and to right cancellation.

Evidently,
$$1 = 1^2 = 1^3 = \ldots = 1^n, \tag{1.8}$$
where n is any positive integer. Since a and a^{-1} commute, we obtain from (1.8) and (1.7) that
$$1^n = 1 = (aa^{-1})^n = a^n(a^{-1})^n.$$
By the uniqueness of the inverse it follows that $(a^{-1})^n$ is the inverse of a^n. It is customary to write
$$(a^n)^{-1} = (a^{-1})^n = a^{-n}, \tag{1.9}$$
and to put
$$a^0 = 1, \tag{1.10}$$
whatever the element a. The reader will have no difficulty in convincing himself that the rules (1.5) and (1.6) are still valid when m and n are any integers, positive, negative or zero. In particular, we observe that two powers of the same element always commute, even when the exponents are negative or zero, thus
$$a^k a^l = a^l a^k. \tag{1.11}$$
If a and b are any two elements we have that
$$(ab)(b^{-1}a^{-1}) = abb^{-1}a^{-1} = 1,$$
whence, by the uniqueness of the inverse,
$$(ab)^{-1} = b^{-1}a^{-1}, \tag{1.12}$$

and more generally

$$(ab\ldots st)^{-1} = t^{-1}s^{-1}\ldots b^{-1}a^{-1}. \tag{1.13}$$

Finally, we remark that 1 is the only **idempotent** element of the group, that is the only solution of the equation

$$x^2 = x \tag{1.14}$$

is $x = 1$.

For on multiplying (1.14) on the left by x^{-1} we obtain that

$$x^{-1}x^2 = x^{-1}x,$$

and hence

$$x = 1.$$

If G consists of a finite number of elements, then this number is called the **order** of G; otherwise G is said to be of infinite order. The order of G, whether finite or infinite, will be denoted by

$$|G|.$$

Although the nomenclature of multiplication is most commonly used for the composition of group elements, it is sometimes convenient to adopt other notations, such as

$$a \circ b$$

to express the composite of a and b.

When the group is Abelian (and in this book only in this case) the additive notation is frequently preferred. Thus we write

$$a+b \ (= b+a)$$

for the composite of a and b. The associative law takes the form

$$(a+b)+c = a+(b+c).$$

The identity (neutral) element is denoted by 0, so that

$$a+0 = 0+a = a,$$

and the inverse is written as $-a$. The analogue of a 'power' of a is now

$$a+a+\ldots+a = na,$$

where on the left n equal terms are involved. It should be noted that the integer n on the right is not usually an element of the group; in fact, *na* is merely an abbreviation for the expression on the left. The 'exponential laws' now take the form

$$(n+m)a = na+ma$$

$$n(ma) = (nm)a,$$

and we introduce the notation

$$-(na) = (-n)a.$$

Since the group is Abelian we have the further relation

$$n(a+b) = na+nb.$$

3. Some Examples of Groups.

Groups abound in most branches of mathematics. We collect here a few examples of groups which the reader will have encountered elsewhere.

(i) *The set of all positive rational numbers form a group with respect to multiplication.* Indeed the product of two positive rational numbers is again a positive rational number, the unit element is the rational number 1, and the inverse of a positive rational number is also such a number. Associativity is known to be one of the laws of arithmetic. This is an infinite Abelian group. Evidently, the set of negative rational numbers does not form a group, neither does the set of positive integers, since each element, other than 1, lacks an inverse.

(ii) *The set of all integers forms an Abelian group with respect to addition.* This group is often denoted by *Z*.

(iii) *Rotations about a fixed point*: if a rigid three-dimensional body is free to move about a fixed point *O*, every displacement of the body is equivalent to a rotation through an angle α about a line *l* passing through *O*. Such a displacement will be denoted by (l, α) or more briefly by a single letter $a = (l, \alpha)$. If *b* is another displacement about *O*, the product *ab* is defined as the displacement which results when *a* is followed by *b*. (In this order—some authors prefer the opposite convention whereby products have to be read from right to left.) Under this law of composition the set of all displacements about *O* forms a non-Abelian group. The identity operation can be expressed as $(l, 0)$, where *l* is arbitrary, and the inverse of (l, α) is $(l, -\alpha)$. The associative law follows from the fact that a rotation is a special kind of linear transformation.

Frequently, we shall be interested only in those displacements which bring the body into coincidence with itself. This subset of displacements also forms a group, which is called the **symmetry group** of the body.

The following illustration demonstrates the fact that the commutative law is not always fulfilled: let 1234 denote a square lamina initially placed in the (x, y)-plane as indicated in Fig. 1, the axis of z being at right angles to the plane of the lamina. We assume that $Oxyz$ is a right-handed system of reference, which is fixed in space.

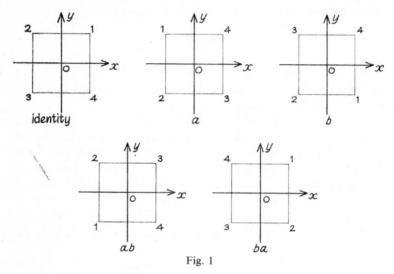

Fig. 1

If, in the above notation,

$$a = (Oz, \tfrac{1}{2}\pi), \quad b = (Ox, \pi),$$

it is seen from the last two diagrams that ab and ba give rise to different positions of the lamina.

(iv) *Groups of matrices:* the reader will be familiar with elementary matrix algebra and in particular with matrix multiplication. Some of the most important examples of groups are provided by certain sets of matrices.

(*a*) Let F be a field, for example the field of real numbers, and consider all non-singular n by n matrices whose elements are chosen arbitrarily from F. This set forms a group under matrix multiplication. It is denoted by $GL(n, F)$ and is called the **general linear group** of degree n over F.

(b) The set of all orthogonal matrices of degree n over F forms a group under matrix multiplication.

(c) The set of non-singular n by n matrices whose elements are integers is closed under multiplication, but the inverse of such a matrix does not in general belong to the set because the formation of the inverse requires division by the determinant. However, the set of integral matrices with determinant ± 1 does form a group; it is called the **unimodular group*** of degree n.

(v) *Residue classes:* let m be a fixed integer greater than one, which in the present context will be referred to as the **modulus**. Two integers x and y are said to be congruent with regard to the modulus m, or *congruent modulo m*, if $x - y$ is divisible by m. This is written symbolically as

$$x \equiv y \pmod{m}, \tag{1.15}$$

and is equivalent to the statement that there exists an integer k such that

$$x = y + km. \tag{1.16}$$

For example, $3 \equiv 18 \pmod 5$, $-2 \equiv 14 \pmod 8$, $12 \equiv 0 \pmod 3$.

Any integer whatever is congruent modulo m with precisely one of the integers in the set

$$Z_m : 0, 1, 2, \ldots, m-2, m-1, \tag{1.17}$$

which is therefore called a complete set of residues modulo m; these are in fact the least non-negative residues modulo m.

It is easy to verify the following rules about congruences:

if $x_1 \equiv y_1 \pmod{m}$ and $x_2 \equiv y_2 \pmod{m}$, then

$$x_1 + x_2 \equiv y_1 + y_2 \pmod{m}. \tag{1.18}$$

and

$$x_1 x_2 \equiv y_1 y_2 \pmod{m}. \tag{1.19}$$

By virtue of (1.18), we can endow (1.17) with an additive group structure by stipulating that $a+b$ shall be that element of (1.17) which is congruent with $a+b$ modulo m. In other words, composition is ordinary addition followed, if necessary, by reduction to the least non-negative residue mod m. The identity element is zero, and the inverse of a is $m-a$. Thus Z_m is a group; it is called the *additive group of residues mod m*. For example, when $m = 5$, $1+2 = 3$, $3+4 = 2$, $2+3 = 0$ and so on.

* Some authors use this term only for matrices of determinant unity.

It may be asked whether (1.19) could be used in a similar manner to introduce a multiplicative group structure into the set of residues. But it would soon become apparent that we should run into difficulties, even if we omit the zero residue, which clearly cannot be an element of a multiplicative group of order greater than one. As we have seen (p. 5), the cancellation law requires that $cx = cy$ implies that $x = y$. But for example, we have that $22 \equiv 4 \pmod 6$, whilst $11 \not\equiv 2 \pmod 6$; thus the cancellation does not in general hold for multiplication modulo m. Nevertheless, as we shall see, cancellation in congruences is permissible in certain cases. In order to analyse the situation, we have to borrow some results and notations from elementary number theory: the highest common factor of a and b is denoted by (a, b); in particular, when $(a, b) = 1$, we say that a and b are **coprime**. If a divides b, we write $a|b$. The following facts are quoted without proof:

(i) if $m|kc$ and $(m, k) = 1$, then $m|c$.

(ii) if $(m, a) = 1$ and $(m, b) = 1$, then $(m, ab) = 1$.

(iii) if $(m, a) = 1$, there exist integers u and v such that $au+mv = 1$.

We can now state that, if $(k, m) = 1$, then the congruence

$$kx \equiv ky \pmod m \tag{1.20}$$

implies that $x \equiv y \pmod m$. For (1.20) is equivalent to $m|k(x-y)$, whence, by (i), $m|x-y$, that is $x \equiv y \pmod m$.

Thus a factor may be cancelled if it is coprime to the modulus.

The numbers of those integers in the set

$$1, 2, \ldots, m$$

which are coprime to m, is denoted by $\phi(m)$ **(Euler's function)**. For example, $\phi(9) = 6$, because there are six integers n such that $1 \leq n \leq 9$ and $(n, 9) = 1$. When p is a prime, all integers in the set $1, 2, \ldots, p$, except the last, are coprime to p, and so

$$\phi(p) = p-1. \tag{1.21}$$

Again, when $m = p^r$, where r is a positive integer, only the multiples of p in the set $1, 2, \ldots, p^r$ fail to be coprime to p; as there are p^{r-1} such multiples, it follows that

$$\phi(p^r) = p^r-p^{r-1}. \tag{1.22}$$

It is customary to put

$$\phi(1) = 1. \tag{1.23}$$

Generally, let

$$R_m: a_1, a_2, \ldots, a_{\phi(m)} \qquad (1.24)$$

be the set of least positive residues which are coprime to m, thus $(a_i, m) = 1$ and $0 < a_i \leqq m$. One of these residues, say a_1, is equal to 1. By (ii), the product of any two elements of (1.24) is again coprime to m; when this product is greater than m, it is not included in (1.24) but is congruent with one of the elements of (1.24), as indeed is any integer which is coprime to m. Thus we may write

$$a_i a_k \equiv a_l \pmod{m}, \qquad (1.25)$$

and define a law of composition in R_m as *multiplication followed, if necessary, by reduction to the least positive residue modulo m*, for example

$$4 \times 5 \equiv 2 \pmod 9, \quad 4 \times 7 \equiv 1 \pmod 9.$$

If it is understood that we are operating in an arithmetic modulo m, so that 'equations' hold only modulo m, it is appropriate to express the law of composition in R_m simply as

$$a_i a_k = a_l. \qquad (1.26)$$

From the properties of congruences it is easy to deduce that the commutative and associative laws are satisfied, and it is clear that $1(= a_1)$ is the unit element. It remains to show that each element $a \in R_m$ possesses an inverse. Since $(a, m) = 1$, we can apply (iii) and deduce the existence of an equation of the form

$$au + mv = 1. \qquad (1.27)$$

This is equivalent to $au \equiv 1 \pmod{m}$. Hence u is the inverse of a in R_m. Thus, under the law of composition specified, R_m forms an Abelian group of order $\phi(m)$.

4. The Multiplication Table. In the abstract theory of groups no reference is made to the nature of the elements. The group is completely given if all possible products ab are known or can be determined by specified rules. In a finite group of order g there are g^2 such products, which may be conveniently listed in a $g \times g$ multiplication table, as was first suggested by A. Cayley.* The following tables display groups of order 2, 3 and 4 respectively

* Phil. Mag., vol. vii (4), 1854.

(i)

	1	a
1	1	a
a	a	1

(ii)

	1	a	b
1	1	a	b
a	a	b	1
b	b	1	a

(iii)

	1	a	b	c,
1	1	a	b	c
a	a	1	c	b
b	b	c	1	a
c	c	b	a	1

(iv)

	1	a	b	c
1	1	a	b	c
a	a	b	c	1
b	b	c	1	a
c	c	1	a	b

In each case the product xy stands in the intersection of the row labelled x and the column labelled y. For example, in (iii) we have that $ac = b$, whilst in (iv) $ac = 1$. The reader will observe that each of these groups happens to be Abelian, which is borne out by the fact that the tables are symmetric about the north-west to south-east diagonal.

A more instructive example is furnished by the group

$$G: 1, a, b, c, d, e \qquad (1.28)$$

of order 6 with the following multiplication table:

(v)

	1	a	b	c	d	e
1	1	a	b	c	d	e
a	a	b	1	e	c	d
b	b	1	a	d	e	c
c	c	d	e	1	a	b
d	d	e	c	b	1	a
e	e	c	d	a	b	1.

$$(1.29)$$

Some group properties are made obvious by the table: the closure is evident because each entry is one of the elements (1.28); the action of the unit element corresponds to the fact that the first row and the first column of the square consist of the elements (1.28) in their original order; the existence of an inverse for each element, and indeed its value, is made plain because precisely one entry in each row and each column is equal to 1. Only the verification of the associative law presents difficulties. To check that $x(yz) = (xy)z$ for all choices of x, y and z would be a laborious task, even for a small group. In the above table the associative law does in fact hold, for example

$$(ac)d = ed = b, \quad a(cd) = a^2 = b,$$

but its general validity is best established by an indirect argument which will be explained presently.

A square table in which each row and column consists of the same elements in some order is sometimes called a **Latin square**. Thus the multiplication table of a finite group is always a Latin square, but the converse is not true, because the associative law may fail. For example the 5×5 Latin square

	1	a	b	c	d
1	1	a	b	c	d
a	a	1	d	b	c
b	b	c	1	d	a
c	c	d	a	1	b
d	d	b	c	a	1

cannot be interpreted as the multiplication table of a group since $(ab)c = dc = a$, whilst $a(bc) = ad = c$ in contradiction to the associative law.

It is readily verified that the set of six matrices

$$\Gamma: \left\{ \begin{array}{lll} I = \begin{bmatrix} 1 & 0 \\ 0 & 1 \end{bmatrix}, & A = \begin{bmatrix} -1 & -1 \\ 1 & 0 \end{bmatrix}, & B = \begin{bmatrix} 0 & 1 \\ -1 & -1 \end{bmatrix}, \\ C = \begin{bmatrix} 0 & 1 \\ 1 & 0 \end{bmatrix}, & D = \begin{bmatrix} 1 & 0 \\ -1 & -1 \end{bmatrix}, & E = \begin{bmatrix} -1 & -1 \\ 0 & 1 \end{bmatrix}. \end{array} \right.$$

$$(1.30)$$

is closed with respect to matrix multiplication; for example

$$B = A^2, A^3 = C^2 = D^2 = E^2 = I, AD = C, AC = E \text{ etc.}$$

Moreover, it will be found that the complete 6 by 6 multiplication table is identical with (1.29) provided that 1 is replaced by I and capital letters are written instead of small letters. Thus if $xy = z$ in accordance with (1.29), the corresponding matrices satisfy the relation $XY = Z$, and conversely any multiplicative relation in Γ is matched by the corresponding relation in G. But matrix multiplication is known to be associative, that is $(XY)Z = X(YZ)$ for any three elements in Γ. Hence we have proved that $(xy)z = x(yz)$ holds in G. We have therefore established the associative law in G. The situation is described by saying that Γ provides a **faithful representation** of G. This is an instance where abstract group theory is being helped by an appeal to a more concrete body of mathematical knowledge.

It is appropriate to say that the groups G and Γ have the same structure. This is an illustration of an important concept which we shall now develop in more detail. Let

$$G: 1, a, b, c, \ldots \tag{1.31}$$

and

$$G: 1', a', b', c', \ldots \tag{1.32}$$

be two (finite or infinite) groups, where the unit elements of G and G' are denoted by 1 and 1' respectively. Suppose there exists a one-to-one correspondence

$$\theta: G \leftrightarrow G' \qquad \textit{bijective map} \tag{1.33}$$

between the elements of G and G', that is to each x in G there is assigned a unique image $x' = x\theta$ in G' and each y' in G' is the image $\leftarrow$ *surjective* of a unique y in G so that $y' = y\theta$. In other words the elements of G and G' have been paired off in such a way that each element of G and G' occurs in precisely one pair. In addition, suppose this correspondence has the property that $xy = z$ if and only if $x'y' = z'$; or, more formally, that

$$(xy)\theta = (x\theta)(y\theta). \tag{1.34}$$

Then we shall say the groups G and G' are **isomorphic** (a Greek expression for 'being of the same shape'), and we write

$$G \cong G'. \tag{1.35}$$

Any relation between the elements of G corresponds to a relation between the elements of G' and conversely; we pass from one group to the other simply by attaching or removing dashes attached to the symbols for the elements. The groups differ merely in notation and, from the abstract point of view, must be regarded as identical, since they have the same multiplication table. Expressed in more technical language we can state that the notion of isomorphism constitutes an *equivalence relation* in the set of all groups. For the usual conditions are clearly satisfied: (i) $G \cong G$ (reflexivity), (ii) if $G \cong G'$, then $G' \cong G$ (symmetry), (iii) if $G \cong G'$ and $G' \cong G''$, then $G \cong G''$ (transitivity). A few more examples will help to elucidate this concept.

Example 1. The following groups of order 4 are isomorphic, the law of composition for each being stated in brackets:

(1) *the numbers* 1, i, -1, $-i$ (*ordinary multiplication*)

(2) *the matrices*   (*matrix multiplication*)

$$\begin{bmatrix} 1 & 0 \\ 0 & 1 \end{bmatrix}, \begin{bmatrix} 0 & 1 \\ -1 & 0 \end{bmatrix}, \begin{bmatrix} -1 & 0 \\ 0 & -1 \end{bmatrix}, \begin{bmatrix} 0 & -1 \\ 1 & 0 \end{bmatrix}$$

(3) *the residues* 1, 2, 4, 3 (*mod* 5)   (*multiplication and reduction modulo* 5)

If the elements in each case are renamed 1, a, b, c, then the multiplication table becomes the table (iv) of p. 12.

Example 2. The following groups of order 4 are isomorphic

(4) *the matrices*   (*matrix multiplication*)

$$\begin{bmatrix} 1 & 0 \\ 0 & 1 \end{bmatrix}, \begin{bmatrix} 1 & 0 \\ 0 & -1 \end{bmatrix}, \begin{bmatrix} -1 & 0 \\ 0 & 1 \end{bmatrix}, \begin{bmatrix} -1 & 0 \\ 0 & -1 \end{bmatrix}$$

(5) *the residues* 1, 3, 5, 7 (*mod* 8)   (*multiplication and reduction modulo* 8)

If the elements in each case are labelled 1, a, b, c the multiplication table is seen to be identical with table (iii) on p. 12.

Obviously, if two finite groups are isomorphic, they must contain the same number of elements. But the converse is not true; for example, the groups given in tables (iii) and (iv) are not isomorphic because in (iii) each element satisfies the equation $x^2 = 1$, which is not the case in table (iv). Thus groups of the same order may have different structures.

5. Cyclic Groups. Consider the set

$$C: 1(= x^0), x, x^{-1}, x^2, x^{-2}, \ldots, x^n, x^{-n}, \ldots \qquad (1.36)$$

of distinct symbols, for which multiplication is defined by the rule

$$x^r x^s = x^{r+s} (r, s = 0, \pm 1, \pm 2, \ldots). \qquad (1.37)$$

Under this law of composition C becomes an Abelian group, which is called the **infinite cyclic group** generated by x. This group is isomorphic with the additive group of integers, that is with the set

$$Z: 0, \pm 1, \pm 2, \ldots$$

in which composition of r and s is defined as $r+s$. The correspondence which establishes the isomorphism is given by

$$x^r \theta = r,$$

where, however, (1.34) has to be replaced by

$$(x^r x^s)\theta = x^r\theta + x^s\theta,$$

since the group Z is written additively. Thus all infinite cyclic groups are isomorphic.

A more interesting situation arises when the symbol x is assumed to satisfy the equation

$$x^m = 1, \tag{1.38}$$

where m is a positive integer greater than unity. In this case the set

$$C_m: 1, x, x^2, \ldots, x^{m-1} \tag{1.39}$$

of distinct symbols forms an Abelian group of order m under the rule

$$x^r x^s = x^{r+s} \ (r, s = 0, 1, \ldots, m-1),$$

where $r+s$ has to be reduced to its least non-negative residue modulo m. This group is called the cyclic group of order m, generated by x. It is isomorphic with the additive group of residues modulo m,

$$Z_m: 0, 1, 2, \ldots, m-1,$$

which was described on p. 11. Hence, again, all cyclic groups of order m are isomorphic. Another representation of the same group is obtained if we replace x in (1.39) by the complex number

$$\varepsilon = \exp(2\pi i/m).$$

Multiplication of any complex number by ε corresponds to a rotation through $2\pi/m$ in the complex plane. If the operation is repeated m times, each point will complete a full cycle, which explains the term cyclic group.

Suppose now that x is an element of any group. Then two cases can arise: either all the powers of x listed in (1.36) are distinct or else there are two integers k and l such that $k > l$ and

$$x^k = x^l$$

and therefore

$$x^{k-l} = 1.$$

Thus in this case a certain positive power of x is equal to the unit element. Hence there must exist a power of least positive exponent having the same property. This remark leads to the following definition.

DEFINITION 4. *Let x be an element of a group. If all the powers of x are distinct, then x is said to be of* **infinite order.** *If not all the powers are distinct, there exists a least positive integer, h, called the* **order (period)** *of x such that*

$$x^h = 1.$$

Of course, in a finite group all elements are of finite order. If x is of order h, then $x^h = 1$ but $x^k \neq 1$ when $0 < k < h$. Again, if $m = hq$, we have that

$$x^m = (x^h)^q = 1.$$

The converse of this remark is also true:

PROPOSITION 1. *If x is of order h, then $x^m = 1$ if and only if m is a multiple of h.*

Proof. Divide m by h, and let q be the quotient and r the remainder so that

$$m = hq + r$$

where $0 \leqq r < h$. Hence

$$1 = x^m = (x^h)^q x^r = 1x^r = x^r.$$

This contradicts the minimal property of h, unless $r = 0$. Hence

$$m = hq.$$

The following facts about the order of an element of a group are easily verified:

(i) *the unit element is the only element of order one.*
(ii) *the elements x and x^{-1} have the same order.*
(iii) *if $y = t^{-1}xt$, were t is any element, then x and y are of the same order.*

PROPOSITION 2. *Let x be of order h. If s is a positive integer, then x^s is of order $h/(h, s)$, where (h, s) denotes the highest common factor of h and s.*

Proof. Let $d = (h, s)$. We then have that

$$h = dh', \ s = ds',$$

where $(h', s') = 1$, and we have to show that x^s is of order h'. Now $(x^s)^{h'} = x^{s'dh'} = (x^{h'd})^{s'} = (x^h)^{s'} = 1$, because x is of order h. It remains to prove that if t is any positive integer such that

$$(x^s)^t = 1, \tag{1.40}$$

then $t \geqq h'$. Suppose that (1.40) is true. Then by Proposition 1, $h|st$, that is $h'd|s'dt$ and hence $h'|s't$. But h' is coprime to s'. Therefore $h'|t$, whence $h' \leqq t$.

6. Maps of Sets. Let $\Sigma: \xi, \eta, \zeta, \ldots$ be a finite or infinite set of objects. A map

$$f: \Sigma \to \Sigma$$

of Σ into itself is a rule whereby to each $\xi \in \Sigma$ there is assigned a unique object $\eta \in \Sigma$, called the image of ξ under f. We write $\eta = \xi f$, in preference to the notation $\eta = f(\xi)$, which is more customary in analysis and topology. Two maps, f and g, are equal if and only if $\xi f = \xi g$ for all $\xi \in \Sigma$. The composite of f and g is the map $f \circ g$ defined by

$$\xi(f \circ g) = (\xi f)g,$$

which means that $f \circ g$ is obtained by letting f be followed by g. Thus if $\xi f = \eta$, then $\xi(f \circ g) = \eta g$.

Let f, g and h be three maps of Σ into itself. We shall show that the composition of these maps always obeys the associative law. Let ξ be any object of Σ and put

$$\xi f = \eta, \quad \eta g = \zeta, \quad \zeta h = \tau.$$

Then

$$\xi[f \circ (g \circ h)] = (\xi f)(g \circ h) = \eta(g \circ h) = (\eta g)h = \zeta h = \tau$$

and

$$\xi[(f \circ g) \circ h] = [\xi(f \circ g)]h = [(\xi f)g]h = (\eta g)h = \zeta h = \tau.$$

Since ξ was an arbitrary object of Σ, it follows that

$$f \circ (g \circ h) = (f \circ g) \circ h. \tag{1.41}$$

Example 1. Let $\Sigma: \xi, \eta, \zeta, \ldots$ be an n-dimensional vector space. We may think of the objects of Σ as row-vectors. If A is an $n \times n$ matrix, then $f: \xi \to \xi A$ is a map of Σ into itself. If $g: \xi \to \xi B$ is another such map, the composite map is given by $f \circ g: \xi \to \xi AB$. Thus the foregoing argument confirms that matrix multiplication is associative.

In order to prove that a collection of maps

$$G: f, g, h, \ldots$$

forms a group it is therefore only necessary to verify the axioms (I), (III) and (IV) of p. 2. We observe that f possesses an inverse if and only if f is one-to-one and maps Σ onto Σ; this means that each $\eta \in \Sigma$ is the image of precisely one object $\xi \in \Sigma$. The relation $\xi f = \eta$ may therefore be uniquely solved for ξ and written as $\xi = \eta f^{-1}$, thus defining the inverse map f^{-1}.

Example 2. The collection of maps which bring a given solid into coincidence with itself evidently satisfies (I), (III) and (IV) and therefore forms a group.

Example 3. Let z range over the extended z-plane, that is over all complex numbers and the point at infinity. The six maps

$$\left. \begin{array}{l} f_1: z \to z \ (\text{identity}), \quad f_2: z \to \dfrac{1}{1-z}, \quad f_3: z \to \dfrac{z-1}{z} \\[3mm] f_4: z \to \dfrac{1}{z}, \quad f_5: z \to 1-z, \quad f_6: z \to \dfrac{z}{z-1} \end{array} \right\} \quad (1.42)$$

transform the extended z-plane into itself and therefore constitute an associative system under composition. It is a remarkable fact that the system is closed. For example

$$z(f_2 \circ f_3) = (zf_2)f_3 = \frac{1}{1-z}f_3 = \frac{(1-z)^{-1}-1}{(1-z)^{-1}} = z = zf_1,$$

so that $f_2 \circ f_3 = f_1$ and hence $f_3 = f_2^{-1}$.

$$z(f_4 \circ f_3) = (zf_4)f_3 = \frac{1}{z}f_3 = \frac{z^{-1}-1}{z^{-1}} = 1-z = zf_5,$$

so that $f_4 \circ f_3 = f_5$, and so on. The complete multiplication table is as follows:

(vi)

	f_1	f_2	f_3	f_4	f_5	f_6
f_1	f_1	f_2	f_3	f_4	f_5	f_6
f_2	f_2	f_3	f_1	f_5	f_6	f_4
f_3	f_3	f_1	f_2	f_6	f_4	f_5
f_4	f_4	f_6	f_5	f_1	f_3	f_2
f_5	f_5	f_4	f_6	f_2	f_1	f_3
f_6	f_6	f_5	f_4	f_3	f_2	f_1

If we write $1, a, b, c, d, e$ in place of $f_1, f_2, f_3, f_4, f_5, f_6$, it will be seen that table (vi) becomes identical with table (v) on p. 12. Thus we have discovered another faithful representation of this abstract group.

7. Permutations. The study of maps which act on a finite set Σ of objects is of especial importance. For simplicity the objects of Σ are often denoted by the integers $1, 2, \ldots, n$. A map of Σ onto itself is called a permutation of **degree** n. It is explicitly described by the symbol

$$\pi = \begin{pmatrix} 1 & 2 & \ldots & j & \ldots & n \\ a_1 & a_2 & \ldots & a_j & \ldots & a_n \end{pmatrix} \tag{1.43}$$

where $a_j = j\pi$ is the image of j under π. Thus the second row in (1.43) is a rearrangement of the integers $1, 2, \ldots, n$. From elementary Algebra it is known that there are $n!$ such rearrangements. Hence there are $n!$ permutations of degree n. The complete set of permutations will be denoted by S_n.

We observe that the information given in (1.43) may be presented in a variety of equivalent ways. In fact we may rearrange the columns in this symbol as we please. For example, the symbols

$$\begin{pmatrix} 1 & 2 & 3 & 4 \\ 2 & 3 & 1 & 4 \end{pmatrix} = \begin{pmatrix} 2 & 1 & 4 & 3 \\ 3 & 2 & 4 & 1 \end{pmatrix} = \begin{pmatrix} 4 & 2 & 1 & 3 \\ 4 & 3 & 2 & 1 \end{pmatrix} = \cdots$$

all denote the same permutation. The first of these, in which the top row contains the objects in their natural order, is called the **standard form.** Evidently, any permutation allows $n!$ equivalent forms, as the top row can be chosen arbitrarily and the relevant information arranged accordingly.

Let

$$\rho = \begin{pmatrix} 1 & 2 & \ldots & n \\ b_1 & b_2 & \ldots & b_n \end{pmatrix} = \begin{pmatrix} a_1 & a_2 & \ldots & a_n \\ c_1 & c_2 & \ldots & c_n \end{pmatrix} \tag{1.44}$$

be another permutation, where $b_j = j\rho$ and $c_j = a_j\rho$. The composition of permutations follows the rule for combining maps. However, for simplicity, we shall write the product as $\pi\rho$ rather than $\pi \circ \rho$. Thus $\pi\rho$ is the permutation that results from first carrying out π and then ρ. Some authors adopt the opposite conventions, which is more appropriate when the image of j under π is denoted by $\pi(j)$ and not by $j\pi$, as we do. When ρ has been 'prepared' for left-multiplication by π, as indicated in (1.44), the product $\pi\rho$ can be written down at once, namely,

$$\pi\rho = \begin{pmatrix} 1 & 2 & \ldots & n \\ c_1 & c_2 & \ldots & c_n \end{pmatrix},$$

because for any j $(j = 1, 2, \ldots, n)$, $j\pi = a_j$ and $a_j\rho = c_j$ so that $j\pi\rho = (j\pi)\rho = a_j\rho = c_j$

For example when

$$\pi = \begin{pmatrix} 1 & 2 & 3 & 4 \\ 2 & 3 & 4 & 1 \end{pmatrix}, \quad \rho = \begin{pmatrix} 1 & 2 & 3 & 4 \\ 3 & 1 & 2 & 4 \end{pmatrix},$$

we find that

$$\pi\rho = \begin{pmatrix} 1 & 2 & 3 & 4 \\ 2 & 3 & 4 & 1 \end{pmatrix}\begin{pmatrix} 2 & 3 & 4 & 1 \\ 1 & 2 & 4 & 3 \end{pmatrix} = \begin{pmatrix} 1 & 2 & 3 & 4 \\ 1 & 2 & 4 & 3 \end{pmatrix}$$

after ρ has been suitably rearranged. Incidentally,

$$\rho\pi = \begin{pmatrix} 1 & 2 & 3 & 4 \\ 3 & 1 & 2 & 4 \end{pmatrix}\begin{pmatrix} 3 & 1 & 2 & 4 \\ 4 & 2 & 3 & 1 \end{pmatrix} = \begin{pmatrix} 1 & 2 & 3 & 4 \\ 4 & 2 & 3 & 1 \end{pmatrix},$$

which demonstrates the fact that multiplication of permutations is, in general, non-commutative.*

The permutation

$$\iota = \begin{pmatrix} 1 & 2 \ldots n \\ 1 & 2 \ldots n \end{pmatrix} = \ldots = \begin{pmatrix} a_1 & a_2 \ldots a_n \\ a_1 & a_2 \ldots a_n \end{pmatrix}$$

which leaves all objects fixed, clearly satisfies the relations $\iota\pi = \pi\iota = \pi$ and is therefore the identity permutation. The inverse of π is given by the symbol

$$\pi^{-1} = \begin{pmatrix} a_1 & a_2 \ldots a_n \\ 1 & 2 \ldots n \end{pmatrix}$$

(in non-standard form); for it is easy to verify that

$$\pi\pi^{-1} = \pi^{-1}\pi = \iota.$$

For example

$$\begin{pmatrix} 1 & 2 & 3 & 4 \\ 2 & 3 & 4 & 1 \end{pmatrix}^{-1} = \begin{pmatrix} 2 & 3 & 4 & 1 \\ 1 & 2 & 3 & 4 \end{pmatrix} = \begin{pmatrix} 1 & 2 & 3 & 4 \\ 4 & 1 & 2 & 3 \end{pmatrix}.$$

There is no need to check the associative law, as this is covered by general properties of maps. We have therefore proved the following theorem.

---

* When the opposite convention for multiplication is adopted, the products $\pi\rho$ and $\rho\pi$ must be interchanged.

THEOREM 1. *The set S_n of all permutations on n objects forms a group of order n!, called the* **symmetric group** *of degree n, the law of composition being that for maps of the objects onto themselves.*

With a little practice the reader will become accustomed to evaluate products of two or more permutations without writing down the intermediate stages of preparation. For example, let

$$\alpha = \begin{pmatrix} 1 & 2 & 3 & 4 \\ 2 & 3 & 1 & 4 \end{pmatrix}, \quad \beta = \begin{pmatrix} 1 & 2 & 3 & 4 \\ 4 & 1 & 2 & 3 \end{pmatrix}, \quad \gamma = \begin{pmatrix} 1 & 2 & 3 & 4 \\ 4 & 3 & 2 & 1 \end{pmatrix}.$$

In order to evaluate the product $\alpha\beta\gamma$ we examine, in turn, the changes which each object undergoes when the operations α, β and γ are carried out in succession. Thus

$$1 \to 2 \to 1 \to 4$$
$$2 \to 3 \to 2 \to 3$$
$$3 \to 1 \to 4 \to 1$$
$$4 \to 4 \to 3 \to 2,$$

where in each row the arrows refer to the actions of α, β and γ (in this order), to be read from left to right.
Hence

$$\alpha\beta\gamma = \begin{pmatrix} 1 & 2 & 3 & 4 \\ 4 & 3 & 1 & 2 \end{pmatrix}.$$

As a further illustration we list the six permutations of S_3:

$$\left. \begin{array}{l} \iota = \begin{pmatrix} 1 & 2 & 3 \\ 1 & 2 & 3 \end{pmatrix}, \quad \alpha = \begin{pmatrix} 1 & 2 & 3 \\ 2 & 3 & 1 \end{pmatrix}, \quad \beta = \begin{pmatrix} 1 & 2 & 3 \\ 3 & 1 & 2 \end{pmatrix} \\[2mm] \gamma = \begin{pmatrix} 1 & 2 & 3 \\ 2 & 1 & 3 \end{pmatrix}, \quad \delta = \begin{pmatrix} 1 & 2 & 3 \\ 3 & 2 & 1 \end{pmatrix}, \quad \varepsilon = \begin{pmatrix} 1 & 2 & 3 \\ 1 & 3 & 2 \end{pmatrix} \end{array} \right\} \quad (1.45)$$

On examining the group structure we realize that S_3 is isomorphic with the abstract group presented in table (v) on p. 12. The isomorphism is established by matching 1 with ι and each Greek letter with the corresponding Latin letter. For example, by the rule for multiplying permutations, we find that

$$\alpha\gamma = \varepsilon, \quad \beta\gamma = \delta,$$

which corresponds to the relations

$$ac = e, \quad bc = d$$

in table (v). Thus we have yet another representation of this abstract group.

Suppose that the set Σ is divided into two mutually disjoint subsets, say

$$\Sigma_1 = \{1, 2, \ldots, m\}, \quad \Sigma_2 = \{m+1, m+2, \ldots, n\}$$

and that π and ρ are permutations of Σ such that π acts only on Σ_1 but leaves each object of Σ_2 unchanged, whilst ρ acts on Σ_2 but does not move any object of Σ_1. Then it is clear that $\pi\rho = \rho\pi$, because the actions of π and ρ do not interfere with each other. Hence we note that permutations which act on mutually exclusive sets of objects, commute with each other.

A permutation which interchanges m objects cyclically is called a cycle of degree m. Thus if the objects are denoted by $1, 2, \ldots, m$, this permutation is described by the symbol

$$\gamma = \begin{pmatrix} 1 & 2 \ldots m-1 & m \\ 2 & 3 \ldots m & 1 \end{pmatrix}. \tag{1.46}$$

If we visualize the m objects arranged at m places on the circumference of a circle, then γ moves each object to the next place so that, in particular, the last object comes to occupy the place of the first. It is customary to write cycles in the contracted notation

$$\gamma = (1 \quad 2 \ldots m),$$

which is to be regarded as equivalent to (1.46). Thus

$$i\gamma = i+1 \quad (i = 1, 2, \ldots, m-1), \quad m\gamma = 1. \tag{1.47}$$

Since it is immaterial with which object the operation begins, we can express γ by any one of the equivalent forms

$$(1 \ 2 \ldots m) = (2 \ 3 \ldots m \ 1) = \ldots (m \ 1 \ldots m-1).$$

The action of γ can be described by the equations (1.47) or more briefly by

$$j\gamma = j+1 \pmod{m}, \tag{1.47'}$$

with the understanding that the right-hand side of (1.47)' is to be reduced to the least positive residue modulo m. Similarly, the effect of the rth power of γ is summarized by

$$j\gamma^r = j+r \pmod{m}. \tag{1.48}$$

It is therefore obvious that $\gamma^m = \iota$, whilst $\gamma^r \neq \iota$ when $0 < r < m$. Thus we find that a cycle of degree m is of order m.

Henceforth we shall adopt the convention that an object which remains fixed under the permutation π need not be explicitly mentioned in the symbol for π. For example, when $n = 3$,

$$(1 \quad 2 \quad 3) = \begin{pmatrix} 1 & 2 & 3 \\ 2 & 3 & 1 \end{pmatrix}$$

and, when $n = 5$,

$$(1 \quad 2 \quad 3) = \begin{pmatrix} 1 & 2 & 3 & 4 & 5 \\ 2 & 3 & 1 & 4 & 5 \end{pmatrix}.$$

Strictly speaking, the symbol (1 2 3) here denotes different permutations, having different degrees, but it is generally clear from the context how many objects are involved and hence which of these objects remain fixed.

It is often convenient to represent a permutation as a product of cycles operating on disjoint sets of objects. Thus, when $n = 7$,

$$\pi = (1 \quad 2) \quad (4 \quad 6 \quad 7)$$

denotes the permutation

$$\pi = \begin{pmatrix} 1 & 2 & 3 & 4 & 5 & 6 & 7 \\ 2 & 1 & 3 & 6 & 5 & 7 & 4 \end{pmatrix}.$$

Any permutation whatever can be resolved into disjoint cycles. In order to see this we introduce the notion of **orbits** under π. Choose any object p and consider what happens to p under repeated applications of π: the set

$$p, p\pi, p\pi^2, \ldots \tag{1.49}$$

is called the orbit of p. As the objects in (1.49) cannot all be distinct, there must be non-negative integers r, s such that $s > r$ and $p\pi^s = p\pi^r$. Hence $p\pi^{s-r} = p$. It follows that there is a least positive integer, h, such that

$$p\pi^h = p. \tag{1.50}$$

It is now clear that π contains the cycle

$$(p, p\pi, p\pi^2, \ldots, p\pi^{h-1}) \tag{1.51}$$

of order h. If q is any object not contained in (1.51), then let k be the least positive integer such that $q\pi^k = q$. Thus q generates the cycle

$$(q, q\pi, q\pi^2, \ldots, q\pi^{k-1}). \tag{1.52}$$

It is important to observe that the cycles (1.51) and (1.52) have no element in common. For suppose that

$$p\pi^a = q\pi^b,$$

and hence

$$q = p\pi^{a-b}.$$

On dividing $a-b$ by h we obtain that

$$a-b = th+r,$$

where $0 \leqq r < h$. Hence it would follow that

$$q = p\pi^r$$

contradicting our choice of q. If there is an object not contained in (1.51) or (1.52), it generates a further cycle whose objects are disjoint from the previous ones, and we continue to construct cycles until all objects are accounted for. An object which remains fixed under π generates a cycle of length one, and this may be omitted according to our convention. Using a more technical term we may say that we have established an equivalence relation between the objects of Σ, two objects being equivalent if and only if they belong to the same orbit and hence to the same cycle. As the reader will know, an equivalence relation on Σ always results in a partition of Σ into disjoint equivalence classes, which in our case correspond to the cyclic factors of π. We have therefore proved the following theorem.

THEOREM 2. *A permutation may be resolved into a product of disjoint cycles. The cycles mutually commute, and the decomposition is unique apart from the rearrangement of the cyclic factors.*

Example. Let

$$\pi = \begin{pmatrix} 1 & 2 & 3 & 4 & 5 & 6 & 7 & 8 \\ 4 & 5 & 6 & 1 & 7 & 8 & 2 & 3 \end{pmatrix}.$$

Starting with the object 1 we find that its orbit is 1, 4. Hence π contains the cycle (1 4). Continuing with any object not in this cycle, say 2, we obtain the orbit 2, 5, 7 giving the cycle (2 5 7). Finally, we reach the orbit 3, 6, 8 and hence the cycle (3 6 8). As there are no more objects to be accounted for, we have shown that

$$\pi = (1\ 4)\quad (2\ 5\ 7)\quad (3\ 6\ 8).$$

Exercises

(1) Prove that the following sets of numbers form infinite Abelian groups with respect to ordinary multiplication:

    (a) $\{2^k\}$ $(k = 0, \pm 1, \pm 2, \ldots)$

    (b) $\left\{\dfrac{1+2m}{1+2n}\right\}$ $(m, n = 0, \pm 1, \pm 2, \ldots)$

    (c) $\{\cos \theta + i \sin \theta\}$, where θ runs over all rational numbers.

(2) Why do the positive rational numbers not form a group when the law of composition for a and b is defined as a/b?

(3) Let a be the map $x \rightarrow \alpha x + \beta$, where α and β are given complex numbers and $\alpha \neq 1$. Obtain a formula for a^n, where n is a positive integer, and show that a is of finite order if and only if α is a root of unity.

In examples (4) to (8) the elements are assumed to lie in a group so that the associative law is taken for granted.

(4) If each of the elements a, b and ab is of order 2, prove that a and b commute.

(5) Prove that the elements ab and ba have the same order.

(6) If $ba = a^m b^n$, prove that the elements $a^m b^{n-2}$, $a^{m-2}b^n$ and ab^{-1} have the same order.

(7) If $b^{-1}ab = a^k$, prove that $b^{-r}a^s b^r = a^{sk^r}$.

(8) Let x be an element of order mn, where $(m, n) = 1$. Prove that x can be expressed in the form $x = yz$, where y and z commute and are of orders m and n respectively.

(9) Prove that a group of even order contains an odd number of elements of order 2.

(10) Find the order of each element in the multiplicative group of residues 1, 2, 3, 4. 5, 6 modulo 7. Show that the group is cyclic of order 6.

(11) Show that the matrices

$$\begin{bmatrix} 1 & 0 \\ 0 & 1 \end{bmatrix},\ \begin{bmatrix} \omega & 0 \\ 0 & \omega^2 \end{bmatrix},\ \begin{bmatrix} \omega^2 & 0 \\ 0 & \omega \end{bmatrix},\ \begin{bmatrix} 0 & 1 \\ 1 & 0 \end{bmatrix},\ \begin{bmatrix} 0 & \omega^2 \\ \omega & 0 \end{bmatrix},\ \begin{bmatrix} 0 & \omega \\ \omega^2 & 0 \end{bmatrix},$$

where $\omega^3 = 1$, $\omega \neq 1$, form a group of order 6 with respect to matrix multiplication. Prove that this group is isomorphic with that given in table (v) on p. 12.

(12) Show that the maps

$$f_1: z \to z, f_2: z \to -z, f_3: z \to \frac{1}{z}, f_4: z \to -\frac{1}{z},$$

of the extended z-plane onto itself form a group, which is isomorphic with the group given in table (iii) on p. 12.

(13) Resolve (i) $\begin{pmatrix} 1 & 2 & 3 & 4 & 5 & 6 & 7 & 8 & 9 \\ 4 & 6 & 9 & 7 & 2 & 5 & 8 & 1 & 3 \end{pmatrix}$ and (ii) $\begin{pmatrix} a & b & c & d & e & f \\ c & e & d & f & b & a \end{pmatrix}$ into mutually disjoint cycles. Find the orders of the two given permutations.

(14) Express in terms of mutually disjoint cycles
   (i) $(abc \ldots k)(al)$;
   (ii) $(a_1 a_2 \ldots a_r x y b_1 b_2 \ldots b_s)(a_r a_{r-1} \ldots a_1 x y c_1 c_2 \ldots c_t)$;
   (iii) $(a_1 a_2 \ldots a_r x y z b_1 b_2 \ldots b_s)(a_r a_{r-1} \ldots a_1 x y z c_1 c_2 \ldots c_t)$.

(15) Verify that the permutations

$$\iota \text{ (identity)}, (12)(34), (13)(24), (14)(23)$$

form an Abelian group of order 4, which is isomorphic with the group given in table (iii) on p. 12.

(16) Show that the set of matrices

$$A(v) = \left(1 - \frac{v^2}{c^2}\right)^{-\frac{1}{2}} \begin{bmatrix} 1 & -v \\ -v/c^2 & 1 \end{bmatrix},$$

where c is a positive constant and v varies in the interval $-c < v < c$, forms a group with the composition law

$$A(v_1)A(v_2) = A(v_3),$$

$$v_3 = \frac{v_1 + v_2}{1 + \dfrac{v_1 + v_2}{c^2}} \qquad \text{(Lorentz Group)}.$$

II. Subgroups

8. Subsets. Since a group G is a collection of elements, the usual definitions and notations of set theory can be applied to G. Thus if $A, B, C, \ldots$ are subsets of G, we write $A \subset B$ to express the fact that every element of A is also an element of B, including the possibility that A and B are equal. The union $A \cup B$ is the set of elements belonging either to A or B or to both, the intersection $A \cap B$ consists of those elements which belong to both A and B; if there is no such element we write $A \cap B = \varnothing$ (the empty set). The notation $a \in A$ means that the element a belongs to A. In some contexts we adopt the (slightly illogical) convention of identifying an element a with the set consisting of the single element a. Thus if the elements of A are $a_1, a_2, a_3, \ldots$ we will write

$$A = a_1 \cup a_2 \cup a_3 \cup \ldots.$$

Now the multiplication defined in G endows the subsets of G with an additional structure. Given any two subsets A and B we define

$$AB \tag{2.1}$$

as the set of all elements that can be expressed in the form ab, where $a \in A$ and $b \in B$. These products need not be distinct, because it may happen that $a_1 \neq a_2$, $b_1 \neq b_2$, but $a_1 b_1 = a_2 b_2$. It should be stressed, however, that AB is merely regarded as a set, so that repetitions of elements are ignored. As usual, subsets are equal if and only if they contain the same distinct elements, irrespective of repetitions. In the sequel equality between subsets is always understood in this sense. Of course, in general

$$AB \neq BA.$$

But even when $AB = BA$, this equality does not mean that each element of A commutes with each element of B. We can infer only that, for any $a \in A$ and $b \in B$, there exist elements $a' \in A$ and $b' \in B$ such that $ab = b'a'$.

28

It is easy to verify that multiplication of subsets is associative, that is

$$(AB)C = A(BC),   \tag{2.2}$$

so that either side of (2.2) may simply be denoted by ABC. Using an obvious abbreviation we put

$$A^2 = AA, \quad A^3 = AAA, \ldots.$$

Thus A^2 is the collection of elements that can be expressed in the form a_1a_2, where a_1 and a_2 range over A.

The following rules are readily established;

$$(A \cup B)C = AC \cup BC,$$
$$C(A \cup B) = CA \cup CB,$$
$$(A \cap B)C = AC \cap BC,$$
$$C(A \cap B) = CA \cap CB.$$

We note the special cases in which some of the sets consist of a single element. Thus, if x and y are elements of G, Ax is the set of all elements of the form ax, and yAx consist of all elements yax, where a ranges over A. We observe that

$$x^{-1}(A_1 \cap A_2 \cap \ldots \cap A_r)x$$
$$= x^{-1}A_1x \cap x^{-1}A_2x \cap \ldots \cap x^{-1}A_rx.   \tag{2.3}$$

When G is an additive Abelian group, composition of two subsets A and B is written as

$$A+B.$$

This is the set of all elements which can be expressed in the form $a+b$, where $a \in A$ and $b \in B$. In particular, the subset

$$A+A$$

(which will not be contracted to $2A$, see p. 6) is the collection of all elements $a+a'$ where a and a' belong to A. When x is a fixed element of G, the subset

$$A+x$$

consists of the elements $a+x$ $(a \in A)$, and we use the notation $A-x$ for $A+(-x)$.

In general, the cancellation rule does not apply to multiplication of subsets, that is, if $AC = BC$, we cannot deduce that $A = B$. But $Ax = Bx$ implies that $A = B$, and $Ax = C$ is equivalent to

$A = Cx^{-1}$. Similar results hold for left multiplication of a set by a single element. An important situation, which we shall encounter in the next chapter, arises when

$$Ax = xA, \quad \text{or} \quad x^{-1}Ax = A. \tag{2.4}$$

This means that for every $a \in A$ there exists an element $a' \in A$ such that $ax = xa'$.

The **cardinal** of A, that is the number of distinct elements in A, whether finite or infinite, will often be denoted by $|A|$.

9. Subgroups. We are particularly interested in those subsets of a group G which obey the group postulates; such subsets are called subgroups of G. Thus H is a subgroup of G if the following conditions are satisfied:

(1) *if $u \in H$ and $v \in H$, then $uv \in H$ (closure)*
(2) *$1 \in H$, where 1 is the identity element of G (identity)*
(3) *if $u \in H$, then $u^{-1} \in H$ (inverse)*.

We did not mention the associative law, because its validity is taken for granted in the whole of G.

When H is a subgroup of G we write

$$H \leqq G$$

in preference to $H \subset G$. The symbol $\leqq$ will be used only for subsets which are groups. Strict inclusion will be denoted by $<$. If H is a subgroup and s one of its elements, then the closure property of H implies that $Hs \subset H$. On the other hand, any element u of H can be written as $u = (us^{-1})s$. Since $us^{-1} \in H$, this shows that $u \in Hs$, and hence $H \subset Hs$. Thus

$$Hs = H \tag{2.5}$$

and similarly

$$sH = H. \tag{2.5}'$$

Conversely, suppose that s is an element of G satisfying (2.5). Then, in particular,

$$s = 1s \in H,$$

so that (2.5) or (2.5)' is a necessary and sufficient condition for an element of G to belong to the subgroup H.

These remarks can easily be generalized, as the reader may verify.

Thus we may state the following result.

PROPOSITION 3. *The subset S of G belongs to the subgroup H if and only if*

$$HS = SH = H.$$

In particular, when $S = H$, we find that

$$H^2 = H. \tag{2.6}$$

It is interesting to observe that, when H is finite, the relation (2.6) conversely implies that H is a group.

PROPOSITION 4. *If H is a finite subset of G, then H is a subgroup if and only if $H^2 = H$.*

Proof. It remains only to prove that (2.6) implies that H is a group. Let the elements of H be enumerated as follows:

$$H: u_1, u_2, \ldots, u_h \tag{2.7}$$

and suppose that u is any one of these elements. Then the h elements

$$Hu: u_1u, u_2u, \ldots, u_hu \tag{2.8}$$

all belong to H^2 and hence, by virtue of our hypothesis, are in H. Moreover, these elements are distinct because the cancellation law holds in G. Hence the sets (2.7) and (2.8) are identical apart from rearrangement. In particular, the element u occurs in (2.8). Thus there exists an integer j such that

$$u_ju = u,$$

whence $u_j = 1 \in H$. Finally, there exists an integer k such that

$$u_ku = 1(= u_j),$$

that is, $u_k = u^{-1} \in H$. This proves that H is a group.

When H is a (finite or infinite) subgroup of G and x is an arbitrary element of G, then the subset

$$H' = x^{-1}Hx \tag{2.9}$$

is also a subgroup of G. For if $x^{-1}ux$ and $x^{-1}vx$ are two arbitrary elements of H', where $u, v \in H$, then $(x^{-1}ux)(x^{-1}vx) = x^{-1}uvx \in H'$. Also $x^{-1}1x = 1 \in H'$ and $x^{-1}u^{-1}x = (x^{-1}ux)^{-1} \in H'$. Moreover, these two groups are isomorphic, since

$$u\theta = x^{-1}ux \quad (u \in H)$$

is a one-to-one mapping of H onto H' with the property that

$$(u\theta)(v\theta) = (uv)\theta.$$

Thus

$$H \cong H'.$$

We remark that every group G has trivially the subgroups $H = \{1\}$ and $H = G$. A subgroup which lies between these two extremes is called a **proper subgroup**.

10. Cosets. Let H be a subgroup of G and let x be any element of G. Then

$$Hx$$

is called a **right coset** of G relative to H, or, more precisely, the right coset generated by x, or the right coset containing x; for it is clear that $x \in Hx$, because $1 \in H$.

When $x = u$, where u is an element of H, then by Proposition 3, $Hu = H$. This shows that H itself is a coset, which alternatively may be written as $H1$ or more generally as Hu, where u is any element of H.

As can be seen from these observations, two distinct elements may generate the same cosets. Let us now find the necessary and sufficient condition that

$$Hx = Hy, \tag{2.10}$$

where x and y are elements of G. If (2.10) is true, then in particular, $x = 1x \in Hy$. Hence there exists an element $u \in H$ such that

$$x = uy,$$

or

$$xy^{-1} \in H. \tag{2.11}$$

Conversely, if (2.11) holds, then

$$Hx = Huy = Hy.$$

Any two cosets are either identical or else have no element in common; in other words, if two cosets have an element in common, then they are identical. For suppose that

$$z \in Hx \cap Hy.$$

Then there exist elements u and v of H such that

$$z = ux = vy.$$

Hence

$$xy^{-1} = u^{-1}v \in H,$$

which implies that $Hx = Hy$. We summarize these results in the following proposition.

PROPOSITION 5. *Let H be a subgroup of a group G. Then the cosets Hx and Hy are identical if and only if $xy^{-1} \in H$. Any two cosets are either identical or else have no element in common.*

It is worthwhile looking at the situation from a more abstract point of view. Let us say that two elements $x, y \in G$ are equivalent (relative to H), written $x \sim y$, if there exists an element $u \in H$ such that $x = uy$, or equivalently if

$$Hx = Hy,$$

that is if x and y lie in the same right coset of H. It is clear that this does, in fact, define an equivalence relation. For (i) $x \sim x$, (ii) $x \sim y$ implies that $y \sim x$, and (iii) if $x \sim y$ and $y \sim z$, then x, y and z all lie in the same coset so that $x \sim z$.

Generally, when an equivalence relation has been defined on a set, this set can be expressed as the union of disjoint subsets, namely the union of all the distinct equivalence classes. In the present case, the equivalence classes are the right cosets so that G is the union of all the distinct cosets. In order to express this result more formally, we select one representative from each coset. If t_i is one of the representatives, the corresponding coset may be denoted by Ht_i. The collection of distinct right cosets may be infinite and even non-enumerable. In that case we use an index set I, whose elements are in one-to-one correspondence with the cosets. The fact that G is the union of all the distinct cosets is then expressed by the formula

$$G = \bigcup_{i \in I} Ht_i. \tag{2.12}$$

The number of distinct right cosets, that is the cardinal of I, is called the **index** of H in G and is denoted by

$$[G : H]. \tag{2.13}$$

When there are infinitely many right cosets we put $[G : H] = \infty$.

The collection $\{t_i, i \in I\}$ of representatives is called a right **transversal** of H in G. Whilst the index is completely determined by

H and G, the transversal is clearly not unique. If one transversal, t_i, is known, the most general transversal is given by

$$\{u_i t_i;\ i \in I\},$$

where the u_i are arbitrary elements of H.

In an analogous manner we can consider **left cosets** of H. A typical left coset is $xH(x \in G)$, and it is readily verified that

$$xH = yH$$

if and only if there exists an element $v \in H$ such that $x = yv$, or alternatively if

$$y^{-1}x \in H. \tag{2.14}$$

As before, two left cosets are either identical or else have no element in common. It follows that G can be split up into the disjoint union of all the distinct left cosets, thus

$$G = \bigcup_{j \in J} s_j H,$$

where J is an index set enumerating the left cosets and $\{s_j\}$ is a set of representatives for left cosets, or, as we shall say a left transversal of H in G. It is, however, easy to see that the index sets I and J have the same cardinal. In fact, starting from the decomposition (2.12) we shall show that

$$G = \bigcup_{i \in I} t_i^{-1} H \tag{2.15}$$

is a decomposition into left cosets. First, we observe that the cosets in (2.15) are distinct. For if

$$t_i^{-1}H = t_k^{-1}H,$$

it would follow that

$$(t_k^{-1})^{-1}t_i^{-1} \in H,\ t_k t_i^{-1} \in H$$

and hence $Ht_k = Ht_i$, which is impossible unless $i = k$. Next, every element $x \in G$ is contained in the union on the right-hand side of (2.15). For x^{-1} must be in one of the right cosets in the decomposition (2.12), say $x^{-1} \in Ht_m$. Hence $x \in t_m^{-1}H$. This establishes (2.15), and we have shown that the left cosets can be enumerated by the same index set as the right cosets. We recall that H is one of the cosets (left or right). When H is finite, as in (2.7), the elements of Ht are

$$u_1t, u_2t, \ldots, u_ht$$

Thus each coset consists of h elements.

We can now prove one of the oldest and most important results about finite groups.

THEOREM 3. (Lagrange): *Let G be a finite group of order g. If H is a subgroup of order h, then*

(i) *h divides g, say $g = nh$, and*

(ii) *n is equal to the index $[G: H]$, so that there exist decompositions*

$$G = \bigcup_{i=1}^{n} Ht_i, \quad G = \bigcup_{i=1}^{n} s_iH \tag{2.16}$$

of G into right and left cosets respectively.

Proof. The existence of the decompositions (2.16) has already been proved in the general case, where n is the index. We only have to show that

$$g = nh.$$

This follows at once by counting the number of elements on both sides of one of the decompositions, since we have seen that each coset contains h element and the disjoint union of n cosets accounts for all g elements of G.

COROLLARY 1. *If G is a finite group of order g, the order of every element is a factor of g. All elements of G satisfy the equation*

$$x^g = 1.$$

Proof. Let u be an element of G. Since G is finite, the order of u is finite, say r. Hence the elements

$$1, u, u^2, \ldots, u^{r-1} \ (u^r = 1)$$

form a cyclic subgroup of order r. By Lagrange's Theorem r divides g so that $g = sr$, where s is a positive integer. Therefore

$$u^g = (u^r)^s = 1.$$

COROLLARY 2. *A group of prime order has no proper subgroups and is necessarily cyclic.*

Proof. Let G be a group of order p, where p is a prime. The order of any subgroup is either one or p, that is the subgroup consists either of the unit element alone or else coincides with G.

If u is any element of G other than 1, then the order of u is greater than one and a factor of p. Hence u is of order p and the elements

$$1, u, u^2, \ldots, u^{p-1}$$

are distinct and are therefore all the elements of G.

Example. In the group of order 6, given in table (v), p. 12, the orders of a, b, c, d, e are 3, 3, 2, 2, 2 respectively.

When G is an additive Abelian group and H is a subgroup of G, a typical coset of H is written

$$H+x$$

and we have that

$$H+x = H+y$$

if and only if

$$x-y \in H,$$

or alternatively

$$x = y+u,$$

where u is an element of H. In this case, we sometimes say that x is *congruent with y modulo H*, and we write

$$x \equiv y(\mathrm{mod}\ H).$$

11. Subgroups of a Cyclic Group. Consider first the case of the infinite cyclic group

$$C: 1(= x^0), x, x^{-1}, x^2, x^{-2}, \ldots \tag{2.17}$$

Disregarding the trivial subgroup $\{1\}$ we may say that every subgroup H of C consists of certain powers of x, including 1, thus

$$H: 1, x^a, x^b, \ldots,$$

where $a, b, \ldots$ are positive or negative integers. Since x^{-a} belongs to H when x^a does, it follows that every non-trivial subgroup of C contains at least one power of x with positive exponent. Hence there exists in H a power of x with least positive exponent, say x^m. Consequently, H contains all elements of the form

$$x^{mq} (q = 0, \pm 1, \pm 2, \ldots), \tag{2.18}$$

that is H contains the cyclic group generated by x^m. We assert that there are no other elements in H besides those listed in (2.18). For let x^a be any element of H. Divide a by m, thus

$$a = mq+r,$$

where $0 \leqq r < m$. Then

$$x^a = x^{mq}x^r,$$

$$x^a x^{-mq} = x^r.$$

Since both factors on the left belong to H, it follows that x^r is an element of H. But this contradicts the minimality of m, unless $r = 0$. Hence $a = mq$, that is (2.18) comprises all elements of H. We see that every non-trivial subgroup of an infinite cyclic group C is itself an infinite cyclic group and is therefore isomorphic with C.

The situation is more interesting when we are concerned with the subgroups of a finite cyclic group. The result is summarized in the following theorem.

THEOREM 4. *Let*

$$C: 1, x, x^2, \ldots, x^{g-1} \ (x^g = 1) \tag{2.19}$$

be a cyclic group order of g. Then corresponding to every divisor h of g there exists one and only one subgroup of order h, which may be generated by $x^{g/h}$.

Proof. (i) Let $g = hn$. The elements

$$1, x^n, x^{2n}, \ldots, x^{(h-1)n} \tag{2.20}$$

are distinct, since an equality between them would lead to a relation

$$x^{ln} = 1,$$

where

$$0 < ln < hn(= g),$$

contradicting the hypothesis that x is of order g. Thus (2.20) forms a subgroup of order h, generated by the element x^n, which is of order h. (ii) Conversely, suppose that $h|g$, say $g = hn$, and that

$$H: 1, u_2, u_3, \ldots, u_{h-1}$$

is a subgroup of order h. Each u_i is a power of x, say

$$u_i = x^{\lambda_i}, \quad (i = 2, 3, \ldots, h-1),$$

where λ_i is an integer satisfying

$$0 < \lambda_i < g.$$

Since H is of order h, Corollary 1 implies that

$$u_i^h = 1,$$

that is

$$x^{h\lambda_i} = 1.$$

By Corollary 1, p. 35, it follows that $g|h\lambda_i$. Hence there exist integers k_i such that

$$h\lambda_i = k_i g = k_i hn$$

$$\lambda_i = k_i n.$$

This proves that each element of H is a power of x^n. Only h of these powers are distinct, and they are listed in (2.20). Hence H is a subset of (2.20), but since H is of order h, it must be identical with the set given in (2.20). This proves that the latter is the unique subgroup of order h.

12. Intersections and Generators. The structure of a group is often illuminated by a study of subgroups. It is therefore important to possess methods for constructing subgroups.

It is obvious that if $H \leqq G$ and $K \leqq H$, then $K \leqq G$. Next, if H and K are subgroups of G, then their intersection

$$D = H \cap K$$

is a subgroup of G. For if x and y belong to D, we have that $x, y \in H$ and $x, y \in K$, whence $xy \in H$, $xy \in K$, that is $xy \in D$; also $1 \in D$, because $1 \in H$ and $1 \in K$; finally if $x \in D$, then $x^{-1} \in H$, $x^{-1} \in K$ and therefore $x^{-1} \in D$. This proves that D is a subgroup. More generally, the intersection of any number of subgroups

$$H \cap K \cap L \ldots$$

is a subgroup.

On the other hand, the union

$$H \cup K$$

of two subgroups is not, in general, a subgroup. For if $u \in H$ and $v \in K$, there is no reason to suppose that uv lies either in H or in K and hence in $H \cup K$. A more elaborate construction is required to obtain the 'smallest' subgroup which contains both H and K.

Let

$$a, b, c, \ldots \tag{2.21}$$

be a collection of elements of G and consider the set of all products consisting of a finite number of factors possibly with repetitions selected from the elements (2.21) or their inverses, for example $a^2b^{-1}cab$; amongst these products we always include the 'empty' product, which we identify with the unit element of G. It is clear that the set of these products forms a group, because if we multiply two products of a finite number of factors we obtain another product of this kind, and the inverse of a product also belongs to this set. The group constructed in this way is denoted by

$$\text{gp } \{a, b, c, \ldots\} = M \tag{2.22}$$

say, and is called the group generated by $a, b, c, \ldots$. Evidently, every group containing the elements (2.21) must contain M, which justifies the statement that M is the smallest subgroup containing these elements. Alternatively, we may say that M is the intersection of all groups containing the elements (2.21). It may, of course, turn out that $M = G$.

The elements $a, b, c, \ldots$ are called **generators** of M. However, it should be pointed out that the generators are not unique nor are they, in general, assumed to be irredundant. For example, the generator a would be redundant if

$$a \in \text{gp } \{b, c, \ldots\},$$

in which case we could replace (2.22) by

$$M = \text{gp } \{b, c, \ldots\}.$$

We are chiefly interested in finitely-generated groups. It is clear that such a group always possesses an irredundant set of generators. In fact we can start with any set of generators and eliminate those which can be expressed in terms of other generators.

Every group G can be expressed in the form (2.22); for example we may trivially regard all elements of G as generators and subsequently remove redundant generators if we wish. For most practical purposes it is desirable to reduce the number of generators as much as possible.

A group with only one generator, x, is the cyclic group generated by x and can now be written as gp $\{x\}$. To illustrate these ideas let us once more turn to the group $G(\cong S_3)$ of order 6 which is exhibited in table (v), p. 12. We find that each of the six elements can be expressed in terms of a and c, thus

$$1 = c^2 (= a^3), a = a, b = a^2, c = c, d = ca, e = ca^2$$
$$\tag{2.23}$$

Hence in this case we can write

$$G = \text{gp} \{a, c\}. \tag{2.24}$$

Alternatively, it can be shown that

$$G = \text{gp} \{b, d\}. \tag{2.25}$$

For a and c, and hence the whole group, can be expressed in terms of b and d, namely

$$a = b^2, \quad c = db.$$

The generators is (2.24) or in (2.25) are certainly irredundant, because the group is non-Abelian and cannot be generated by a single element, in which case it would be cyclic and hence Abelian.

It is important to realize that irredundant generators may nevertheless be linked by non-trivial relations. Thus by referring to table (v) we find that

$$ac = ca^2, \tag{2.26}$$

which is equivalent to

$$(ac)^2 = 1, \tag{2.26}'$$

because $(ac)^2 = acac = acca^2 = a1a^2 = a^3 = 1$. It is impossible to solve either of these equations so as to express one of the generators in terms of the other. An equation like (2.26) or (2.26)' is called a defining relation. It is often convenient to specify a particular group by a set of generators and a set of defining relations. We shall return to this principle in more detail later (Chapter 5) but mention here that in the present case the equations

$$a^3 = c^2 = (ac)^2 = 1 \tag{2.27}$$

serve as a set of defining relations. Indeed, the information contained in (2.27) suffices to construct the entire multiplication table. First, we remark that the six elements

$$1, a, a^2, c, ca, ca^2 \tag{2.28}$$

are certainly distinct; for example if a were equal to ca^2, it would follow that $a^{-1} = c$, contradicting the fact that a and c are irredundant generators. Next, we check that the system (2.28) is closed under multiplication by virtue of the relations (2.27); for example

$$(ca)(ca^2) = c(ac)a^2 = cca^2a^2 = c^2a^4 = a,$$
$$a^2c = a(ac) = aca^2 = ca^4 = ca,$$

and so on, a factor c being systematically moved to the left by virtue of (2.26), until the product is identical with one of the elements (2.28). In this notation the complete multiplication table is as follows:

(viii)

	1	a	a^2	c	ca	ca^2
1	1	a	a^2	c	ca	ca^2
a	a	a^2	1	ca^2	c	ca
a^2	a^2	1	a	ca	ca^2	c
c	c	ca	ca^2	1	a	a^2
ca	ca	ca^2	c	a^2	1	a
ca^2	ca^2	c	ca	a	a^2	1

and this provides yet another version of the group first presented in table (v), p. 12. If $A, B, C, \ldots$ are subsets of a group G, the group generated by them is denoted by

$$\text{gp} \{A, B, C, \ldots\}$$

and is defined as the collection of all finite products in which each factor is an element of A or B or $C \ldots$ or the inverse of such an element, taken in any order with or without repetition. This reduces to the previous concept of generators if we regard as generators all the elements of $A \cup B \cup C \cup \ldots$ We might equally well write instead

$$\text{gp} \{A \cup B \cup C \cup \ldots\}.$$

Of course, if A is a subgroup, we have that $A = \text{gp} \{A\}$.

13. The Direct Product. We shall now discuss a simple method of constructing a new group out of two given groups. Let H and K be any groups and consider the set of all pairs

$$(u, v),$$

where u and v range over H and K respectively. The set of these pairs is denoted by

$$G = H \times K$$

and is called the **(exterior) direct product** of H and K. The set G is turned into a group by endowing it with the composition law

$$(u, v)(u', v') = (uu', vv'). \tag{2.29}$$

It is easy to verify that the associative law holds in G, because multiplication is associative in H and in K. The unit element in G is the pair

$$(1_H, 1_K),$$

where 1_H and 1_K are the unit elements of H and K respectively. Also

$$(u, v)^{-1} = (u^{-1}, v^{-1}).$$

If H and K are finite groups of orders h and k respectively, then $H \times K$ is a group of order hk.

More generally, if $H_1, H_2, \ldots, H_r$ are any groups their direct product

$$H_1 \times H_2 \times \ldots \times H_r$$

consists of all r-tuples

$$(u_1, u_2, \ldots, u_r),$$

where $u_i \in H_i (i = 1, 2, \ldots, r)$ and multiplication is carried out in each component of the r-tuple. If each H_i is finite, then clearly

$$|H_1 \times H_2 \times \ldots \times H_r| = \prod_{i=1}^{r} |H_i|.$$

It sometimes happens that a group G is isomorphic with the direct product of two of its subgroups, H and K, thus

$$G \cong H \times K, \tag{2.30}$$

or with a slight abuse of notation

$$G = H \times K. \tag{2.30'}$$

This situation arises in the following circumstances:

(1) The subgroups H and K commute elementwise, that is, if u and v are arbitrary elements of H and K respectively, then

$$uv = vu. \tag{2.31}$$

(2) Every element $x \in G$ can be expressed in the form $x = uv$, or more briefly

$$G = HK. \tag{2.32}$$

(3) The intersection of H and K is the unit element, that is

$$H \cap K = 1. \tag{2.33}$$

We remark that (2) and (3) are equivalent to the single condition:
(2′) Every element $x \in G$ can be *uniquely* decomposed as $x = uv$, where $u \in H$ and $v \in K$.

For assume (2) and (3) and suppose we have two decompositions

$$x = uv = u_1v_1.$$

Then

$$u_1^{-1}u = v_1v^{-1}, \tag{2.34}$$

and the element on either side of (2.34) belongs to H and to K, whence, by (3), it must be equal to 1. Thus $u = u_1$ and $v = v_1$, which proves the uniqueness of the decomposition, as required in (2').

Conversely, assume (2') and suppose that $w \in H \cap K$. Then $w = 1w = w1$ are two decompositions of w with factors in H and K respectively, and we deduce from the uniqueness that $w = 1$.

The foregoing conditions make it plain that each element $x \in G$ uniquely determines a pair (u, v) where $u \in H$ and $v \in K$ and that each such pair occurs, because (u, v) corresponds to the product $uv = x$. The correspondence

$$x\theta = (uv)\theta = (u, v)$$

establishes the isomorphism (2.30), because, by (1)

$$(uv)(u'v')\theta = (uu'\,vv')\theta = (uu', vv').$$

Similarly, we have that

$$G \cong H_1 \times H_2 \times \ldots \times H_r,$$

where $H_i(i = 1, 2, \ldots, r)$ are subgroups of G, if the following conditions are satisfied:

(1) Any two groups H_i, H_j commute elementwise.

(2) Every element x of G can be expressed in the form

$$x = u_1u_2 \ldots u_r, \tag{2.35}$$

where $u_i \in H_i$

(3) $$H_i \cap H_1H_2 \ldots H_{i-1}H_{i+1} \ldots H_r = \{1\},$$

or, alternatively, in place of (2) and (3),

(2') The decomposition (2.35) is unique. In particular, if

$$u_1u_2 \ldots u_r = 1,$$

it follows from (2') that

$$u_1 = u_2 = \ldots = u_r = 1,$$

because $1 = 11 \ldots 1$ is the only decomposition of 1.

When a group has been expressed as the direct product of sub-groups we speak of an **interior direct product.**

Example. The least positive residues prime to the modulus 15 are

$$1, 2, 4, 7, 8, 11, 13, 14. \tag{2.36}$$

They form an Abelian group of order 8 (see p. 50), which, as we shall see, is isomorphic with the direct product of the cyclic groups generated by the elements 2 and 11 respectively. In fact, the residue 2 generates a cyclic group of order 4, namely

$$C_4: 1, 2, 4, 8 \quad (2^4 = 16 \equiv 1 \pmod{15}).$$

Similarly, 11 generates a cyclic group of order 2,

$$C_2: 1, 11 \quad (11^2 = 121 \equiv 1 \pmod{15}).$$

Since the whole group is Abelian, we have only to check conditions (2) and (3) of p. 42. Taking all possible products we obtain

$$1, 2, 4, 8, 11, 22, 44, 88,$$

which on reduction mod. 15 becomes

$$1, 2, 4, 8, 11, 7, 14, 13.$$

As this is the complete group, condition (2) holds, and we see at once that

$$C_4 \cap C_2 = \{1\}.$$

This shows that the group is isomorphic with $C_4 \times C_2$.

The following simple proposition, which is of some independent interest, will be used in the next section.

PROPOSITION 6. *Let G be a finite group in which all elements satisfy the equation*

$$x^2 = 1, \tag{2.37}$$

that is each element, other than the unit element, is of order 2. *Then G is isomorphic with an Abelian group of the type*

$$C_2 \times C_2 \times \ldots \times C_2,$$

and the order of G is therefore a power of 2.

Proof. By Corollary 2, p. 35, the proposition is obviously true when G is the (only) group of order 2. Suppose then that G is of order greater than 2 and let a and b be distinct elements other than 1. By hypothesis

$$a^2 = b^2 = 1,$$

so that

$$a = a^{-1}, b = b^{-1}.$$

Next, consider the element ab. By (2.37), $(ab)^2 = 1$, whence

$$ab = (ab)^{-1} = b^{-1}a^{-1} = ba.$$

This shows that G is Abelian. Let $u_1, u_2, \ldots, u_r$ be a set of irredundant generators of G. Since G is Abelian, products of the generators can be collected in such a way that every element can be expressed in the 'normal' form

$$u_1^{k_1}u_2^{k_2} \ldots u_r^{k_r}. \tag{2.38}$$

But by virtue of (2.37) the exponents in (2.28) can be restricted to take only the values 0 and 1. Then all the products will be distinct; for an equality between two such products would lead to a relation

$$u_1^{l_1}u_2^{l_2} \ldots u_r^{l_r} = 1,$$

where each l_i is either zero or unity. This would enable us to express one of the generators in terms of the others, contradicting the hypothesis that the generators were irredundant. Thus the normal form (2.38) is unique, which amounts to saying that

$$G = \text{gp } \{u_1\} \times \text{gp } \{u_2\} \times \ldots \times \text{gp } \{u_r\},$$

and hence

$$G \cong C_2 \times C_2 \times \ldots \times C_2$$

(r factors).

14. Survey of Groups up to Order 8. No successful method has yet been discovered for constructing all possible abstract groups of preassigned order, nor do we know in advance how many such groups exist, except in a few simple cases.

However, the elementary means which we have so far developed, suffice to give a complete list of groups up to order 8. Since groups of prime order have already been discussed (Corollary 2, p. 35), it remains only to discuss in more detail the cases in which the order, g, is equal to

$$4 \text{ or } 6 \text{ or } 8.$$

There are two groups of order 4, both Abelian.

For if $g = 4$, an element, other than 1, can only be of order 4 or 2 (Corollary 1, p. 35).

(1) If G contains an element a of order 4, this element generates G; in fact, the four elements of G are

$$1, a, a^2, a^3 \ (a^4 = 1)$$

and we have that $G = C_4$, the cyclic group of order 4.

(2) Next, suppose that G has no element of order 4. Then all elements, other than 1, are of order 2 and we deduce from Proposition 6 that

$$G = C_2 \times C_2.$$

Thus G is generated by two elements a and b, and the four elements of G are

$$1, a, b, ab, \tag{2.39}$$

where

$$a^2 = b^2 = 1, \ ab = ba \tag{2.40}$$

This group is called the **four-group** (F. Klein's '*Vierergruppe*') and is often denoted by V.

There being no other possibilities we conclude that any group of order 4 is isomorphic either with C_4 or with $V(\cong C_2 \times C_2)$. In a different notation the multiplication tables of these groups were presented in (iii) and (iv) on p. 12.

There are two groups of order 6, one cyclic and one non-Abelian.

(1) If G possesses an element a of order 6, then

$$G = \text{gp} \{a\} = C_6.$$

(2) Next, suppose there is no element of order 6. The order of every element, other than 1, is therefore 2 or 3 (Corollary 1, p. 35). Since the order of G is not a power of 2, not all its elements can satisfy (2.37). Hence there exists at least one element, a, of order 3, so that

$$1, a, a^2 \tag{2.41}$$

are three distinct elements of G and

$$a^3 = 1, \tag{2.42}$$

If c is a further element of G, the six elements

$$1, a, a^2, c, ca, ca^2 \tag{2.43}$$

are distinct, as we pointed out on p. 40, in connection with the elements listed in (2.28).

If the elements (2.43) are to form a group of order 6, the axiom of closure must be fulfilled. In particular, c^2 must be one of these elements. We cannot have an equation of the form $c^2 = ca^i (i = 0, 1, 2)$, as this would imply that c belongs to the set (2.41). There remain only the following three possibilities:

$$(\alpha) \ c^2 = 1, \quad (\beta) \ c^2 = a, \quad (\gamma) \ c^2 = a^2. \tag{2.44}$$

Under the assumptions (β) and (γ), the elements c cannot be of order 2 and must therefore be of order 3. But on multiplying (β) and (γ) on the left by c we should obtain $1 = ca$ and $1 = ca^2$ respectively, neither of which can be true. Hence we conclude that (α) must hold, that is

$$c^2 = 1. \tag{2.45}$$

Next, consider ac. It must be amongst the elements (2.43). As it cannot be equal to c or to a power of a, we are left with the alternatives

$$ac = ca, \quad \text{or} \quad ac = ca^2. \tag{2.46}$$

The first of these renders the group Abelian. Let us find the order of ac in this case. Thus

$$(ac)^2 = a^2c^2 = a^2 \neq 1, \ (ac)^3 = a^3c^3 = c^3 = c \neq 1,$$

and the element ac would have to be of order 6, contrary to our initial assumption. Hence the second equation of (2.46) must hold that is

$$ac = ca^2, \quad \text{or, equivalently,} \quad (ac)^2 = 1,$$

see (2.26) and (2.26)'. To summarize we can state that if there is a group G of order 6 other C_6, then

$$G = \text{gp} \{a, c\}$$

subject to the relations

$$a^3 = c^2 = (ac)^2 = 1.$$

This does not prove that such a group exists. But we happen to know that this is the case; indeed its multiplication table is exhibited on p. 41. Thus there are precisely two groups of order 6.

There are five groups of order 8, of which three are Abelian and two are non-Abelian.

Three Abelian groups of order 8 are easily written down, namely:

(1) $C_8 = \text{gp}\{a\}$, where $a^8 = 1$ (Table (viii), p. 50).

(2) $C_4 \times C_2 = \text{gp}\{a\} \times \text{gp}\{b\}$, where $a^4 = b^2 = 1$, $ab = ba$ (Table (ix), p. 50).

(3) $C_2 \times C_2 \times C_2 = \text{gp}\{a\} \times \text{gp}\{b\} \times \text{gp}\{c\}$, where $a^2 = b^2 = c^2 = 1$, $ab = ba$, $bc = cb$, $ca = ac$ (Table (x), p. 51).

From the general theory, which we shall develop in Chapter IV, it will follow that these are all possible Abelian groups of order 8, but we shall here derive this result from first principles. If the group contains an element of order 8, it must be the group C_8, and if all elements, other than 1, are of order 2, then it is isomorphic with the group (3).

Henceforth we shall assume that each element, other than 1, is either of order 4 or of order 2 and that there is at least one element of order 4, say a, where

$$a^4 = 1, \quad a^2 \neq 1. \tag{2.47}$$

If b is an element not contained in gp $\{a\}$, then the eight elements

$$1, a, a^2, a^3, b, ab, a^2b, a^3b \tag{2.48}$$

are distinct and therefore constitute the whole group, if such a group exists.

Now b^2 must be one of these elements and in fact must be one of the first four, since b is not a power of a. The equations $b^2 = a$ or $b^2 = a^3$ have to be ruled out, since they would imply that the order of b is eight. Thus there remain the possibilities

$$(\alpha)\ b^2 = 1 \quad \text{or} \quad (\beta)\ b^2 = a^2 \tag{2.49}$$

(α): Assume that $b^2 = 1$. The product ba must be one of the last three elements in (2.48).

(α, i): If $ba = ab$, the group is Abelian and is the group listed under (2).

(α, ii): If $ba = a^2b$, we could deduce that $b^{-1}a^2b = a$, whence

$$(b^{-1}a^2b)^2 = b^{-1}a^4b = b^{-1}1b = 1 = a^2,$$

which is impossible. Hence we must conclude that

(α, iii): $ba = a^3b$,   or, equivalently, $(ab)^2 = 1$.

The group defined by the relations

$$a^4 = b^2 = (ab)^2 = 1 \qquad (2.50)$$

does in fact exist. It is denoted by D_4 and is called the **dihedral group** of order 8 (Table (xi), p. 51). It belongs to a class of groups which will be discussed later (p. 51), when the associative law will be confirmed (see also exercise 7, p. 55).

(β). Assume that $b^2 = a^2$. In this case both a and b are of order 4. Again, ba must be one of the last three elements of (2.48), which we shall consider in turn:

(β, i). If $ba = ab$, the group is Abelian. The element $c = ab^{-1}$ is of order 2 and since $b = c^{-1}$ is of order 2 and since $b = c^{-1}a$, the generator b may be replaced by c. The eight elements may therefore be written as in (2.48), but with c in place of b. Once again we arrive at the group (2).

(β, ii). The relation $ba = a^2b$ is impossible, as it would imply that $ba = b^2b$, that is $a = b^2$, which is inadmissible.

(β, iii). The only remaining alternative, namely $ba = a^3b$, is feasible and, as we shall see, leads to a group defined by the relations

$$a^4 = 1, \ a^2 = b^2, \ ba = a^3b. \qquad (2.51)$$

In order to demonstrate that such a group does in fact exist, we construct a faithful matrix representation. Let

$$A = \begin{bmatrix} \sqrt{-1} & 0 \\ 0 & -\sqrt{-1} \end{bmatrix}, \quad B = \begin{bmatrix} 0 & 1 \\ 1 & 0 \end{bmatrix}.$$

The reader will have no difficulty in verifying that these matrices satisfy the relations (2.51) with the appropriate change of notation and that the eight matrices

$$I, \ A, \ A^2, \ A^3, \ B, \ AB, \ A^2B, \ A^3B$$

are distinct and therefore constitute a multiplicative matrix group, which is isomorphic with the group envisaged under (β, iii).

This group is known as the **quaternion group** (Table (xii), p. 51). We recall that a quaternion is a hypercomplex number.

$$a_0 1 + a_1 i + a_2 j + a_3 k,$$

where the coefficients a_0, a_1, a_2, a_3 are real numbers and the symbols

$$1, \ i, \ j, \ k$$

satisfy the relations

$$i^2 = j^2 = -1, ij = -ji = k$$

or, equivalently,

$$i^4 = 1, i^2 = j^2, ji = i^3j,$$

which agrees with (2.51) apart from the notation.

To summarize our discussion of groups of order 8 we append the complete multiplication tables of the five possible abstract groups of this order:

$$C_8 = \text{gp}\,\{a\}, a^8 = 1$$

	1	a	a^2	a^3	a^4	a^5	a^6	a^7
1	1	a	a^2	a^3	a^4	a^5	a^6	a^7
a	a	a^2	a^3	a^4	a^5	a^6	a^7	1
a^2	a^2	a^3	a^4	a^5	a^6	a^7	1	a
a^3	a^3	a^4	a^5	a^6	a^7	1	a	a^2
a^4	a^4	a^5	a^6	a^7	1	a	a^2	a^3
a^5	a^5	a^6	a^7	1	a	a^2	a^3	a^4
a^6	a^6	a^7	1	a	a^2	a^3	a^4	a^5
a^7	a^7	1	a	a^2	a^3	a^4	a^5	a^6

[Table (viii)]

$$C_4 \times C_2 = \text{gp}\,\{a\} \times \text{gp}\,\{b\} \times \text{gp}\,\{b\}, a^4 = b^2 = 1.$$

	1	a	a^2	a^3	b	ab	a^2b	a^3b
1	1	a	a^2	a^3	b	ab	a^2b	a^3b
a	a	a^2	a^3	1	ab	a^2b	a^3b	b
a^2	a^2	a^3	1	a	a^2b	a^3b	b	ab
a^3	a^3	1	a	a^2	a^3b	b	ab	a^2b
b	b	ab	a^2b	a^3b	1	a	a^2	a^3
ab	ab	a^2b	a^3b	b	a	a^2	a^3	1
a^2b	a^2b	a^3b	b	ab	a^2	a^3	1	a
a^3b	a^3b	b	ab	a^2b	a^3	1	a	a^2

[Table (ix)]

$C_2 \times C_2 \times C_2 = \text{gp}\,\{a\} \times \text{gp}\,\{b\} \times \text{gp}\,\{c\}$, $a^2 = b^2 = c^2 = 1$

	1	a	b	c	ab	ac	bc	abc
1	1	a	b	c	ab	ac	bc	abc
a	a	1	ab	ac	b	c	abc	bc
b	b	ab	1	bc	a	abc	c	ac
c	c	ac	bc	1	abc	a	b	ab
ab	ab	b	a	abc	1	bc	ac	c
ac	ac	c	abc	a	bc	1	ab	b
bc	bc	abc	c	b	ac	ab	1	a
abc	abc	bc	ac	ab	c	b	a	1

[Table (x)]

Dihedral group: $a^4 = b^2 = (ab)^2 = 1$

	1	a	a^2	a^3	b	ab	a^2b	a^3b
1	1	a	a^2	a^3	b	ab	a^2b	a^3b
a	a	a^2	a^3	1	ab	a^2b	a^3b	b
a^2	a^2	a^3	1	a	a^2b	a^3b	b	ab
a^3	a^3	1	a	a^2	a^3b	b	ab	a^2b
b	b	a^3b	a^2b	ab	1	a^3	a^2	a
ab	ab	b	a^3b	a^2b	a	1	a^3	a^2
a^2b	a^2b	ab	b	a^3b	a^2	a	1	a^3
a^3b	a^3b	a^2b	ab	b	a^3	a^2	a	1

[Table (xi)]

Quaternion group: $a^4 = 1$, $a^2 = b$, $ba = a^3b$

	1	a	a^2	a^3	b	ab	a^2b	a^3b
1	1	a	a^2	a^3	b	ab	a^2b	a^3b
a	a	a^2	a^3	1	ab	a^2b	a^3b	b
a^2	a^2	a^3	1	a	a^2b	a^3b	b	ab
a^3	a^3	1	a	a^2	a^3b	b	ab	a^2b
b	b	a^3b	a^2b	ab	a^2	a	1	a^3
ab	ab	b	a^3b	a^2b	a^3	a^2	a	1
a^2b	a^2b	ab	b	a^3b	1	a^3	a^2	a
a^3b	a^3b	a^2b	ab	b	a	1	a^3	a^2

[Table (xii)]

15. The Product Theorem. At the beginning of this chapter we defined the product of two subsets. We shall now examine the case in which both these subsets are subgroups of a group. It will appear that the product of two subgroups is not always a subgroup, but that explicit information can be obtained in the finite case about the number of elements in the product.

THEOREM 5 (Product Theorem). (i) *Let A and B be subgroups. Then the subset AB is a group if and only if*

$$AB = BA. \tag{2.52}$$

(ii) *In the case of finite subgroups, let* $|A| = a, |B| = b, |A \cap B| = d$. *Then, irrespective of whether* (2.52) *holds,*

$$|AB| = |BA| = ab/d.$$

Proof. (i) Since A and B are groups we have that $A^2 = A$ and $B^2 = B$. First, suppose that (2.52) is satisfied and put $H = AB$. Then

$$H^2 = ABAB = A^2B^2 = AB = H,$$

which proves the closure of H. Obviously $1 \in H$, since $1 \in A$ and $1 \in B$. Finally if a and b are arbitrary elements of A and B respectively, then $b^{-1}a^{-1} \in BA$; and hence, by (2.52), $b^{-1}a^{-1} \in AB = H$; that is $(ab)^{-1} \in H$, which completes the proof that H is a group.

Conversely, suppose that $H = AB$ is a group. Hence if a and b are arbitrary elements of A and B respectively, $ab \in H$, $a^{-1}b^{-1} \in H$, and also $(a^{-1}b^{-1})^{-1} \in H$, that is $ba \in H$. This means that

$$BA \subset AB.$$

In particular, $b^{-1}a^{-1} = a_1b_1$, where a_1 and b_1 are certain elements of A and B respectively. Hence $(b^{-1}a^{-1})^{-1} = ab = b_1^{-1}a_1^{-1}$, that is

$$AB \subset BA.$$

Hence we conclude that $AB = BA$.

(ii) Let $D = A \cap B$. Since D is a subgroup of B, we may decompose B into cosets with respect to D, say

$$B = Dt_1 \cup Dt_2 \cup \ldots \cup Dt_n, \tag{2.53}$$

where

$$Dt_i \neq Dt_j, \quad \text{if} \quad i \neq j, \tag{2.54}$$

and

$$n = b/d. \tag{2.55}$$

Multiplying (2.53) on the left by A and observing that $AD = A$ because $D \subset A$, we obtain that

$$AB = At_1 \cup At_2 \cup \ldots \cup At_n. \tag{2.56}$$

We claim that no two of the cosets on the right of (2.56) have an element in common; for if not, we should have an equation of the form

$$u_1 t_i = u_2 t_j,$$

where $u_1, u_2 \in A$ and $i \neq j$. Thus

$$t_i t_j^{-1} = u_1^{-1} u_2.$$

Now the element on the left belongs to B by (2.53), and the element on the right lies in A. Hence either side denotes an element of D. But $t_i t_j^{-1} \in D$ implies that $Dt_i = Dt_j$, which contradicts (2.54). Thus the cosets in (2.56) are disjoint, and since each consists of a elements, we have that

$$|AB| = ad = ab/d.$$

The argument is clearly symmetric in A and B, so that also $|BA| = ab/d$.

16. Double Coset. We saw on p. 33 that the decomposition of a group into cosets relative to a subgroup may be regarded as an instance of dividing a set into equivalence classes with respect to a suitably defined equivalence relation.

Following Frobenius we shall now discuss a different equivalence relation, which involves two subgroups. Let A and B be subgroups of G, which need not be distinct, and term two elements $x, y \in G$ equivalent, written $x \sim y$, if there exist elements $u \in A$ and $v \in B$ such that

$$y = uxv. \tag{2.57}$$

It is easy to check that this is an equivalence relation on G. For

   (i) $x \sim x$, since we may take $u = 1$, $v = 1$,

   (ii) if $x \sim y$, then $y \sim x$, because (2.57) implies that $x = u^{-1} y v^{-1}$,

   (iii) if $x \sim y$, $y \sim z$, that is $y = uxv$, $z = u'yv'$, where $u' \in A$, $v' \in B$, then $z = (u'u)x(vv')$, so that $x \sim z$.

Thus the set G may be divided into the disjoint equivalence classes which stem from this definition of equivalence. The equivalence class containing x is the complex AxB, which is called a double coset of G with respect to A and B. We choose a representative from each class and obtain the decomposition

$$G = \bigcup_{i \in I} At_i B, \tag{2.58}$$

where I is a, possibly infinite, index set in one-to-one correspondence with the set of double cosets. It is clear that (2.58) is a generalization of the left or right coset decomposition, as is seen by taking A or B to be the trivial group $\{1\}$. In contrast to single coset decompositions the double cosets in (2.58) are not, in general, of the same cardinal.

We will pursue the matter further when G is a finite group. Let $|G| = g$, $|A| = a$ and $|B| = b$. First, we observe that the complexes $At_i B$ and $(t_i^{-1} At_i)B$ have the same cardinal, because their elements may be put into one-to-one correspondence by matching $ut_i v$ with $t_i^{-1}(ut_i v)$. Thus

$$|At_i B| = |(t_i^{-1} At_i)B|.$$

Now $t_i^{-1} At_i$ is a subgroup (see p. 31), and

$$|t_i^{-1} At_i| = |A| = a.$$

On applying Theorem 5 to the subgroups $t_i^{-1} At_i$ and B we find that

$$|At_i B| = ab/d_i,$$

where $d_i = |t_i^{-1} At_i \cap B|$. Collecting these results we obtain the following theorem.

THEOREM 6. (Frobenius). *Let G be a finite group of order g and let A and B be subgroups of orders a and b respectively. Then there exist elements $t_1, t_2 \ldots t_r$ such that G is the disjoint union of double cosets, namely*

$$G = At_1 B \cup At_2 B \cup \ldots \cup At_r B.$$

The number of elements in $At_i B$ is ab/d_i, where

$$d_i = |t_i^{-1} At_i \cap B|,$$

and consequently

$$g = ab \sum_{i=1}^{r} d_i^{-1}. \tag{2.59}$$

Exercises

(1) Let $D = X \cap Y$ and $M = \text{gp}\,\{X, Y\}$, where X and Y are non-empty subsets of a group G. Show that, if Z is another subset of G,
$$X \cap Y \cap Z = D \cap Z, \quad \text{gp}\,\{X, Y, Z\} = \text{gp}\,\{M, Z\}.$$

(2) Let $D = A \cap B$, where A and B are subgroups of G. Prove that if $u, v \in At \cap Bs$, where $s, t \in G$, then $Du = Dv$. Deduce that, when $[G:A]$ and $[G:B]$ are finite, $[G:D] \leq [G:A][G:B]$. (*Theorem of Poincaré.*)

(3) Prove that if A and B are finite subgroups whose orders are relatively prime, then $A \cap B$ consists only of the unit element.

(4) Prove that a finite group of composite order has a proper subgroup.

(5) Find all subgroups of order 4 of the dihedral group of order 8 (Table (xi), p. 51).

(6) Show that the group of Table (v) (p. 12) may be defined by the relations
$$c^2 = d^2 = (cd)^3 = 1.$$

(7) Let $\varepsilon = \exp\,(2\pi i/n)$, where n is a positive integer greater than one. Prove that the matrices
$$A = \begin{bmatrix} \varepsilon & 0 \\ 0 & 1/\varepsilon \end{bmatrix}, \quad B = \begin{bmatrix} 0 & 1 \\ 1 & 0 \end{bmatrix}$$
form a faithful representation of the dihedral group $D_n = \text{gp}\,\{a, b\}$ of order $2n$, given by the defining relations
$$a^n = b^2 = (ab)^2 = 1.$$

(8) Let $\theta = \exp\,(\pi i/m)$, where m is a positive integer greater than one. Prove that the matrices
$$A = \begin{bmatrix} \theta & 0 \\ 0 & 1/\theta \end{bmatrix}, \quad B = \begin{bmatrix} 0 & 1 \\ -1 & 1 \end{bmatrix}$$
form a faithful representation of the **dicyclic group** of order $4m$, given by the relations
$$a^{2m} = 1, \quad b^2 = (ab)^2 = a^m.$$

(9) Prove that if a non-commutative group of order 12 contains an element of order 6, then it is isomorphic either with the dihedral group or else with the dicyclic group of that order.

(10) Show that the residues prime to 21 form a (multiplicative) Abelian group which is isomorphic with $C_6 \times C_2$.

(11) Let $X = \text{gp}\,\{x\}$, the infinite cyclic group generated by x, and let $R = \text{gp}\,\{x^r\}$, where r is a positive integer. Prove that $[X:R] = r$.

III. Normal Subgroups

17. Conjugacy Classes. On p. 33 we discussed a method of dividing a group G into equivalence classes relative to a subgroup of G. We shall now introduce an equivalence relation of a different kind.

DEFINITION 4. *The elements a and b are said to be* **conjugate** *in G, if there exists an element t in G such that*

$$b = t^{-1}at. \tag{3.1}$$

We say that t transforms a into b; at this stage we are not particularly interested in the element that accomplishes the transformation, and it should be observed that, for given a and b, there may well be more than one element t satisfying (3.1). The right hand side of (3.1) is sometimes abbreviated to a^t, that is we put

$$a^t = t^{-1}at. \tag{3.2}$$

For a moment let us write $b \sim a$, if there exists a relation (3.1) with some t. We verify that it is an equivalence relation, thus:

(i) $a \sim a$ (reflexive property), taking $t = 1$;

(ii) $b \sim a$ implies $a \sim b$ (symmetric property); for if (3.1) holds, then $a = tbt^{-1} = (t^{-1})^{-1}bt^{-1}$, so that $a \sim b$.

(iii) if $a \sim b$ and $b \sim c$, then $a \sim c$ (transitive property); for we are given that $a = t^{-1}bt$ and $b = s^{-1}cs$, whence on eliminating b, $a = (st)^{-1}c(st)$, that is $a \sim c$.

Furthermore, we note that conjugation obeys the important multiplicative rule

$$(xy)^t = x^t y^t, \tag{3.3}$$

where x, y, t are arbitrary elements of G, for

$$(xy)^t = t^{-1}xyt = (t^{-1}xt)(t^{-1}yt) = x^t y^t.$$

Clearly, (3.3) can be generalized to any number of factors, thus

$$(x_1 x_2 \ldots x_n)^t = x_1{}^t x_2{}^t \ldots x_n{}^t.$$

By putting $y = x^{-1}$ in (3.3) and noting that $1^t = 1$, we deduce that

$$1 = x^t (x^{-1})^t,$$

that is

$$(x^t)^{-1} = (x^{-1})^t. \tag{3.3}'$$

We recall that, when an equivalence relation is defined on a set, this set is broken up into disjoint classes. Each class consists of all those elements which are equivalent to a particular element. In the present case, we speak of **conjugacy classes**. The conjugacy class which contains a particular element a will be denoted by (a); it contains all those elements of G which are conjugate with a, including a itself, thus

$$(a) = t_1^{-1}at_1 \cup t_2^{-1}at_2 \cup \ldots$$

and we may assume that $t_1 = 1$. If b is an element not contained in (a), then b generates a new conjugacy class

$$(b) = s_1^{-1}bs \cup s_2^{-1}bs_2 \cup \ldots,$$

and the transitivity of the conjugacy relation ensures that (a) and (b) have no element in common. Proceeding in this manner we obtain the decomposition of G into conjugacy classes, thus

$$G = (a) \cup (b) \cup (c) \cup \ldots$$

We call $a, b, c, \ldots$ representatives of the various classes, but it must be borne in mind that these representatives are not uniquely determined; indeed $(a) = (a')$, if and only if $a' = x^{-1}ax \ (x \in G)$.

When G is infinite, there may be infinitely many conjugacy classes, and a particular conjugacy class may contain an infinity of elements. It is important to obtain more precise information about the elements which constitute a given class and to determine the size of the class when it is finite. Obviously, the class (1) consists of the single element 1, because $t^{-1}1t = 1$ for all t in G.

In order to examine this problem in more detail we introduce the notion of the **centralizer**. Let a be a fixed element of G and denote by $C(a)$ the set of all those elements of G which commute with a. Thus

$$C(a) = \{t \in G \,|\, ta = at\}.$$

It is easy to verify that $C(a)$ is a subgroup of G. For (i) if $s, t \in C(a)$, then $a(st) = sat = (st)a$ so that $st \in C(a)$; (ii) $1 \in C(a)$, and (iii) if $t \in C(a)$, then $t^{-1}a = at^{-1}$, that is $t^{-1} \in C(a)$.

Incidentally, unless $G = \{1\}$, $|C(a)| \geq 2$; for if $a = 1$, then $C(a) = G$; and if $a \neq 1$, then $a \in C(a)$ and $1 \in C(a)$.

c

Next, consider the coset decomposition of G relative to $C(a)$, say

$$G = \bigcup_i C(a)t_i \quad (i \in I),$$

where I is a suitable index set. We claim that the cosets of $C(a)$ are in one-to-one correspondence with the elements if (a); this correspondence is set up by the map

$$\theta : C(a)x \to x^{-1}ax. \tag{3.4}$$

First we must show that θ is well-defined, bearing in mind that $C(a)x$ could equally be written as $C(a)ux$, where u is an arbitrary element of $C(a)$. Thus we have to demonstrate that the substitution of ux for x does not alter the right-hand side of (3.4). Indeed,

$$(ux)^{-1}a(ux) = x^{-1}u^{-1}aux = x^{-1}ax,$$

because $u \in C(a)$. Since x is any element of G, the map θ is clearly a surjection on to the class (a). Finally, we observe that θ is injective; for if $x^{-1}ax = y^{-1}ay$, then $xy^{-1} \in C(a)$, whence $C(a)x = C(a)y$. Hence θ is bijective, as asserted.

We collect these results in the following:

PROPOSITION 7. *Let a be an element of G and C(a) its centralizer. Then the elements of the conjugacy class are in one-to-one correspondence with the cosets of C(a) in G. In particular when C(a) is of finite index, then $|(a)| = [G:C(a)]$.*

COROLLARY. *If G is a finite group of order g and if h_a is the number of elements in (a), then $h_a | g$.*

Proof. Let $|C(a)| = c_a$ then by Proposition 7, $h_a = g/c_a$, that is $g = c_a h_a$.

Suppose the finite group G has k distinct conjugacy classes. Let $a_1(= 1), a_2, \ldots, a_k$ be a set of representatives and put $h_i = |(a_i)|$. Then

$$G = (a_1) \cup (a_2) \cup \ldots \cup (a_k).$$

whence, on counting elements on each side of this equation,

$$g = h_1 + h_2 + \ldots + h_k. \tag{3.5}$$

This relation is called the **class equation** of G.

18. The Centre. The set Z of elements which commute with every element of G is called the **centre** of G. Thus

$$Z = \{z | tz = zt, \text{ for all } t \in G\}.$$

This is a subgroup of G; for (i) if $tz_1 = z_1t$ and $tz_2 = z_2t$, then $tz_1z_2 = z_1tz_2 = z_1z_2t$ so that $z_1z_2 \in Z$; (ii) if $tz = zt$, then $z^{-1}t = tz^{-1}$ so that $z^{-1} \in Z$; (iii) $1 \in Z$. Of course Z is always Abelian, but it might be the unit group; for example, in the non-Abelian group of order 6 (see table (v), p. 12) the centre consists of the unit element alone. Clearly, $G = Z$ if and only if G is Abelian.

An element of the centre is characterized by the fact that it forms a conjugacy class by itself, for if z is conjugate only with itself then $t^{-1}zt = z$ for all $t \in G$, which means that $z \in Z$. For this reason a central element is sometimes called **self-conjugate**. The following result is interesting because it establishes the existence of a non-trivial centre for an important class of groups.

THEOREM 7. *If G is finite group such that $|G| = p^m$, where p is a prime and $m > 0$, then the centre of G has order p^μ, where $0 < \mu \leqq m$.*

Proof. In the present case the class equation (3.5) becomes

$$p^m = h_1 + h_2 + \ldots + h_k, \tag{3.6}$$

where $h_\alpha | p^m$ ($\alpha = 1, 2, \ldots, k$). Since p is a prime, this implies that each h_α is either equal to unity or else to a power of p. We already know that $h_1 = 1$. Suppose that there are precisely $l (\geqq 1)$ values of α such that $h_\alpha = 1$. We can then write (3.6) in the form

$$p^m = l + ps,$$

for some integer s. It follows that l is divisible by p and since l is positive, we infer that $l \geqq p$. Thus there are at least p self-conjugate elements, that is Z is non-trivial. Since Z is a subgroup of G, Lagrange's theorem furnishes the further information that $|Z| = p^\mu$ where $0 < \mu \leqq m$.

19. Normal Subgroups. The notion of centralizer may be extended to any non-empty subset A of G. Thus the centralizer $C(A)$ of A consists of all those elements of G which commute with each element of A, that is

$$C(A) = \{t | ta = at \quad \text{for all} \quad a \in A\}.$$

As before, the centralizer is always a subgroup of G, possibly the unit subgroup. It is, in fact, the intersection of the groups $C(a)$, where a ranges over A. When it is desirable to express the dependence on the containing group G, we write more precisely $C_G(A)$. Note that, generally,

$$C_G(G) = \text{centre of } G.$$

We now pass to a different concept of commutativity: given a non-empty subset A we consider those elements s of G which satisfy

$$sA = As \tag{3.7}$$

as a relation between subsets. Thus (3.7) means that for every $a \in A$, there exist $a_1, a_2 \in A$ such that $sa = a_1s$ and $as = sa_2$. We leave to the reader the straightforward verification of the fact that those elements s which satisfy (3.7), form a subgroup of G. This subgroup is called the **normalizer** of A and is denoted by

$$N(A), \text{ or, more precisely, by } N_G(A).$$

Evidently, every element of $C(A)$ will certainly satisfy (3.7), as it commutes 'elementwise' with the elements of A. Thus

$$C(A) \leqq N(A).$$

But in general, the normalizer is larger than the centralizer.

We are particularly interested in the case in which A is a subgroup, H, of G. As we have seen (p. 31), when x is any element of x, then $H' = x^{-1}Hx$ is also a subgroup, which is isomorphic with H, though in general distinct from it. We call H and H' **conjugate**. However, conjugation by distinct elements x and y may produce the same subgroup. In fact, the equation

$$x^{-1}Hx = y^{-1}Hy \tag{3.8}$$

is equivalent to $Hxy^{-1} = xy^{-1}H$, and so $xy^{-1} \in N(H)$. Thus (3.8) holds if and only if $x = sy$, where $s \in N(H)$. The group H is counted among its conjugates, and we have that $H = s^{-1}Hs$ if and only if $s \in N(H)$.

Evidently $H \leqq N(H)$, because if $u \in H$, then $u^{-1}Hu = H$ (see Proposition 3, p. 31).

By far the most interesting subgroups of G are those for which the normalizer is the whole of G. When $N(H) = G$, we say that H is a **normal** or **invariant** subgroup of G and we use the special symbol

$$H \lhd G.$$

Let us dwell a little longer on this crucial concept. A normal subgroup is characterized by the fact that it possesses no conjugate subgroups apart from itself, that is

$$xH = Hx, \quad \text{or} \quad H = x^{-1}Hx \quad \text{for all} \quad x \in G. \qquad (3.9)$$

More explicitly, if x is any element of G and u is an element of a normal subgroup H, then there exists an element $u' \in H$ such that

$$x^{-1}ux = u'.$$

Actually, in order to show that $H \lhd G$, it is sufficient to verify that

$$x^{-1}Hx \subset H \qquad (3.10)$$

for all $x \in G$. For if this is the case, we can replace x by x^{-1} and obtain that $xHx^{-1} \subset H$, which is equivalent to

$$H \subset x^{-1}Hx.$$

This, together with (3.10) implies that $H = x^{-1}Hx$. In every group G, the unit subgroup $\{1\}$ and G itself are normal subgroups of G. A group of order greater than unity is called **simple** if it has no normal subgroups other than these trivial ones. Of course, a group of prime order is necessarily simple, but there are interesting examples of simple groups of composite order (see p. 139). In an Abelian group every subgroup is automatically normal, because (3.9) is a consequence of commutativity.

In order to test whether H is normal in G, one of the following procedure is often useful:

(1) Suppose that G is given in terms of generators (see p. 39), say

$$G = \text{gp} \{a, b, c, \ldots\}.$$

If we are able to show that

$$a^{-1}Ha = H, \; b^{-1}Hb = H, \; c^{-1}Hc = H, \; \ldots$$

then $a, b, c, \ldots$ belong to $N(H)$. But since these elements generate the whole of G, we deduce that $G = N(H)$, that is $H \lhd G$.

(2) If H is given in terms of generators, say

$$H = \text{gp} \{x_1, x_2, x_3, \ldots\},$$

then $H \lhd G$, if for every $t \in G$

$$x_i^{\,t} \in H \quad (i = 1, 2, \ldots).$$

For by (3.3), this ensures that every (finite) product of the x's remains in H under conjugation by t, and so $t^{-1}Ht \subset H$.

For example, let G be the dihedral group of order 8 displayed in Table (xi) (p. 51), and let

$$H = \text{gp } \{a\},$$

the cyclic group of order 4 generated by a. Obviously, $a \in N(H)$. Also $bab^{-1} = a^3$, and hence $bHb^{-1} \leqq H$; but the groups bHb^{-1} and H are isomorphic and therefore of the same order. Therefore $bHb^{-1} = H$. This means that $b(= b^{-1})$ belongs to $N(H)$, which proves that $N(H) = G$.

We shall now collect a few elementary facts about normal subgroups.

(i) *The centre is always a normal subgroup*; for the condition (3.9), namely $x^{-1}Zx = Z$ is certainly satisfied for all $x \in G$. Indeed, we even have that $x^{-1}zx = z$ for each element z of Z.

(ii) *If $N_1, N_2, \ldots, N_r$ are normal subgroups, so is their intersection*; for since $x^{-1}N_ix = N_i$ $(i = 1, 2, \ldots, r)$ we deduce that

$$x^{-1}(N_1 \cap N_2 \cap \ldots \cap N_r)x = N_1 \cap N_2 \cap \ldots \cap N_r.$$

(iii) *A subgroup H is normal in G if and only if it is the union of complete conjugacy classes of G*, that is

$$H = (1) \cup (u) \cup (v) \cup \ldots. \tag{3.11}$$

For (3.11) is evidently equivalent to the statement that whenever w belongs to H, so does $x^{-1}wx$, where x is any element of G. This means that $x^{-1}Hx \subset H$, and hence $H \lhd G$.

(iv) *If H is of index 2 in G, then $H \lhd G$.* In this case there are precisely two cosets of H in G, one of which is H and the other consists of $G \backslash H$, that is those elements of G which do not belong to H. Thus if $t \in G \backslash H$, then $G \backslash H = Ht$; the same argument can be used for left cosets, so that $G \backslash H = tH$, whence $Ht = tH$ if $t \notin H$. On the other hand, if $w \in H$, then $H = Hw = wH$. Hence the equation $xH = Hx$ holds for all $x \in G$, that is $H \lhd G$. For example, this method would prove at once that gp $\{a\}$ is a normal subgroup of the dihedral group.

20. Quotient (Factor) groups. The paramount importance of normal subgroups rests on the fact that the collection of cosets of a normal subgroup can be endowed with a group structure. Suppose that

$H \lhd G$ and consider the product (as subsets) of two cosets Hx and Hy. Since $xH = Hx$ and $H^2 = H$, we find that

$$HxHy = HHxy = Hxy. \qquad (3.12)$$

Thus the product of two cosets is again a coset. It is crucial to observe that (3.12) is a genuine relation between cosets, which is independent of the representatives. More precisely, we claim that if $Hx = Hx'$ and $Hy = Hy'$, then $Hxy = Hx'y'$. Indeed, our hypothesis implies that $x' = ux$, $y' = vy$, where $u, v \in H$. It follows that $x'y' = uxvy = uv'xy$, where v' is a suitable element of H. Hence we have that $Hx'y' = Hxy$ as required.

The matter may be put in a slightly different manner if we use the notion of equivalence relative to H. As in Chapter II, p. 33, let us write $x \sim x'$, if there exists an element $u \in H$ such that $x' = ux$. Since $Hx = xH$, we could, alternatively, have stipulated that $x' = xu'$, where $u' \in H$. The equivalence class $[x]$ of a particular element $x \in G$ is therefore identical with the coset $Hx(= xH)$, and (3.12) expresses a multiplication for equivalence classes, namely

$$[x][y] = [xy], \qquad (3.13)$$

which is independent of the class representatives.

By (3.12), the collection of cosets is closed under multiplication of subsets. This raises the hope that the cosets will, in fact, form a group. There is no difficulty about the associative law, as this holds for all subsets (see p. 29). For multiplication of cosets, the unit element is the group H, regarded as one of the cosets, because

$$H(Ht) = (Ht)H = Ht.$$

Finally, the inverse of Ht is the coset Ht^{-1}, since $(Ht)(Ht^{-1}) = H = (Ht^{-1})(Ht)$. The group of cosets we have constructed is denoted by G/H and is called the **quotient** (or **factor**) group of G by H. The order of G/H is equal to the index of H in G, that is

$$|G/H| = [G : H]. \qquad (3.14)$$

The notion of a quotient group is fundamental for group theory and indeed is one of the most important concepts in mathematics. We therefore repeat some of the relevant points:

(1) The elements of G/H are the distinct cosets of H, the law of composition being multiplication of subsets (or addition of cosets when G is written additively, see p. 36).

(2) The identity (neutral) element is the group H, regarded as one of the cosets.

(3) It is immaterial whether we use right or left cosets, since $Ht = tH$ because H is normal.

(4) Recall that the representative of a particular coset is not unique (see p. 32).

(5) The term quotient group and the notation G/H will be used only when H is a normal subgroup.

We shall now illustrate the concept of a quotient group by discussing a few examples:

(i) Let Z be the (additive) group of all integers and let $m > 1$ be a fixed integer. Then the set

$$H: 0, \pm m, \pm 2m, ..., \pm km, ...$$

forms a subgroup of Z. Since Z is Abelian, this is a normal subgroup. If x is any integer, we can write $x = qm+r$, where $0 \leq r < m$. Since qm lies in H, x lies in the coset $H+r$ (see p. 36). Allowing for the possible values of r, we see that

$$H(= H+0), H+1, H+2, \ldots, H+(m-1) \qquad (3.15)$$

are the distinct cosets, that is the elements of Z/H. The cosets are in one-to-one correspondence with the elements of Z_m (see (1.17)). If, for a moment, the elements of Z_m are denoted by $\bar{0}, \bar{1}, \ldots \overline{m-1}$, we can express the correspondence formally as

$$H+r \leftrightarrow \bar{r}.$$

We now observe that the law of composition is preserved by this correspondence. For

$$(H+r)+(H+s) = H+t,$$

where $t \equiv r+s \pmod{m}$ and $0 \leq t < m$, just as

$$\bar{r}+\bar{s} = \bar{t}$$

in accordance with the rules of composition in Z_m. Hence we conclude that

$$Z/H \cong Z_m.$$

(ii) In the quaternion group (Table (xii), p. 51) the element $a^2 = b^2$ obviously commutes with a and with b and hence with every element, because a and b generate the whole group. Hence

$$H = 1 \cup a^2 \quad (a^4 = 1)$$

is a normal subgroup (actually the centre). The elements of G/H can be listed as

$$H, Ha, Hb, Hab. \qquad (3.16)$$

For we know in advance that there are $[G:H] = 8/2 = 4$ cosets, and the cosets (3.16) are distinct, as can easily be verified; for example $Hb = b \cup a^2b$. Now G/H is a group of order 4 and must therefore be isomorphic either with C_4 or with $C_2 \times C_2$ (see p. 46). The question is settled by observing that the square of each element of G/H is equal to the unit element H; in fact $(Ha)^2 = Ha^2 = H$, because $a^2 \in H$; similarly $(Hb)^2 = H$. Finally, since G/H is necessarily Abelian, being of order 4, we have that

$$(Hab)^2 = (Ha)^2(Hb)^2 = H.$$

Hence $G/H \cong C_2 \times C_2$ (see also Proposition 6, p. 44).

(iii) Let $G = GL(n, F)$ be the general linear group of degree n over F (see example (iv), p. 8), that is the set of all non-singular $n \times n$ matrices $a = (a_{ij})$ with entries from F. Then the matrices of determinant unity form a subgroup U. For if $\det u = \det v = 1$, then $\det (uv) = 1$; also $\det u^{-1} = 1$ and the identity matrix belongs to U. Moreover, $U \lhd G$, because if $x \in G$, then $\det (x^{-1}ux) = \det u = 1$. Now it is easy to see that two matrices a and b belong to the same coset of U if and only if $\det a = \det b$; for this is equivalent to $ab^{-1} \in U$, because $\det(ab^{-1}) = 1$. Clearly, the determinant can assume any non-zero value of F. The set of non-zero elements of F is often denoted by $F^\times$. Thus,

$$G/U \cong F^\times.$$

A transversal for U in G (see p. 33) is, for example, furnished by the diagonal matrices diag $(d, 1, 1, \ldots, 1)$, where d ranges over $F^\times$.

Finally, we mention a result about the centre which is sometimes useful.

PROPOSITION 8. *If G is non-Abelian with centre Z, then G/Z is never cyclic.*

Proof. If G/Z were a cyclic group, then all cosets of Z could be expressed as Zt^i, where t is a suitable element of G not contained in Z and $i = 0, \pm 1, \pm 2, \ldots$. Now if x and y are arbitrary elements of G belonging to the cosets Zt^k and Zt^l respectively, we should have that

$$x = z_1t^k, \quad y = z_2t^l,$$

where $z_1, z_2 \in Z$. Hence

$$xy = z_1 t^k z_2 t^l = z_1 z_2 t^{k+l} = yx,$$

that is G would be Abelian, contrary to our hypothesis.

COROLLARY: *A group of order p^2, where p is a prime, is necessarily Abelian.*

Proof. By Theorem 7, $|Z|$ is equal to p or p^2, If $|Z| = p^2$, then $G = Z$, and the group is Abelian. If not, then $|Z| = p$ and $|G/Z| = p$. Hence G/Z would be cyclic, which is ruled out by the preceding Proposition.

21. Homomorphism. The structure of a group consists of the law by which all possible products ab are evaluated. In Chapter I (p. 14) we discussed the situation in which two groups are isomorphic, that is, have the same structure. We shall now consider a more general relationship, in which groups have 'similar' structures; or to use the Greek term, we shall be concerned with **homomorphic** groups. To make this concept more precise, suppose we have a map

$$\theta: G \rightarrow G'$$

of a group G into a group G'. As before, the image of $x \in G$ is denoted by $x\theta$, so that $x' = x\theta$ is a unique element of G' associated with x by the map θ. We say that θ is a homomorphic map, or more briefly, a homomorphism of G into G' if, for all $x, y \in G$,

$$(x\theta)(y\theta) = (xy)\theta. \tag{3.17}$$

This is the sole condition we impose on θ. The following points should be carefully noted:

(i) Let 1 and $1'$ be the identity elements of G and G' respectively. By taking $x = y = 1$ we have that $(1\theta)^2 = 1\theta$. Thus 1θ is an idempotent element of G' and hence (p. 6)

$$1\theta = 1', \tag{3.18}$$

that is, every homomorphism maps the identity element of G onto the identity element of G'. Furthermore, if we let $y = x^{-1}$, we deduce from (3.17) that

$$x^{-1}\theta = (x\theta)^{-1}. \tag{3.19}$$

(ii) We do not demand that the map is one-to-one, thus it may happen that $x_1\theta = x_2\theta$, whilst $x_1 \neq x_2$. If, however, the equation

$x_1\theta = x_2\theta$ always implies that $x_1 = x_2$, then we say that θ is a **monomorphism** or that θ is **injective**.

(iii) In general, it is not assumed that θ is surjective. In other words, there may be elements of G' which are not images of elements of G. The set of images is denoted by $\mathrm{im}\theta$ or, more conveniently, by $G\theta$; it is easy to see that $G\theta$ is a subgroup of G' (which may coincide with G'); indeed if $x', y' \in G\theta$, there exist elements $x, y \in G$ such that $x' = x\theta, y' = y\theta$. Hence $x'y' = (xy)\theta \in G\theta$. Also by (3.18), $1' \in G\theta$ and $(x')^{-1} \in G\theta$ if $x' \in G\theta$. If, on the other hand, θ is surjective, that is if

$$G\theta = G', \tag{3.20}$$

we call θ an **epimorphism**. An isomorphism (in the previous sense) is characterized by the fact that it is both injective and surjective, or more briefly, that it is **bijective**. In this case we continue to use the notation $G \cong G'$.

We shall now establish the important fact that every homomorphic map of G is associated with a normal subgroup of G. Let K be the set of those elements of G which are mapped into $1'$. This set is called the **kernel** of θ, often written as $\ker \theta$. First, we shall show that K is a subgroup of G: if $u, v \in K$, then $u\theta = v\theta = 1'$ and hence, by (3.17), $(uv)\theta = 1'$, also $1 \in K$ by (3.18), and $u^{-1} \in K$ by (3.19). Moreover, K is a normal subgroup, for if $x \in G$ and $u \in K$, then

$$(x^{-1}ux)\theta = (x\theta^{-1}(u\theta)(x\theta) = (x\theta)^{-1}1'(x\theta) = 1',$$

which means that $x^{-1}ux \in K$. Thus we have verified that (3.10) holds for the group K, that is

$$K \lhd G. \tag{3.21}$$

Of course, it may happen that K is the unit subgroup of G. In this connection it is useful to note the following result.

PROPOSITION 9. *The homomorphism θ is injective if and only if $\ker \theta$ consists of the identity element alone.*

Proof. Suppose that θ is injective and let $u \in \ker \theta$. Then $1\theta = u\theta = 1'$. Hence $u = 1$ because θ is injective. Conversely, assume that $\ker \theta = \{1\}$, and suppose that $x\theta = y\theta$. Then $(xy^{-1})\theta = (x\theta)(y\theta)^{-1} = 1'$. Thus $xy^{-1} \in \ker \theta$ and therefore $xy^{-1} = 1$, that is $x = y$, which proves that θ is injective.

Returning to the general case we are now in the position to establish one of the most important facts of group theory.

THEOREM 8. (First* Isomorphism Theorem). *Let $\theta: G \to G'$ be a homomorphism of G into G', with image group $G\theta$ and kernel K. Then*

$$G/K \cong G\theta. \tag{3.22}$$

Proof. We have to produce a bijective homomorphism between the two groups mentioned in (3.22). This is accomplished by a map ϕ, which is related to θ or 'induced' by θ, though in general distinct from it. We recall that the elements of G/K are the cosets Kx, whilst the elements of $G\theta$ are of the form $x\theta$, where $x \in G$, and all elements of Kx have the same image. It is therefore tempting to suggest that there is a one-to-one correspondence, ϕ, based on the definition

$$(Kx)\phi = x\theta. \tag{3.23}$$

Although this will turn out to be correct, the definition (3.23) is unacceptable without further justification. For we know that the generating element x of Kx is not uniquely determined (Proposition 5, p. 33), and we have to convince ourselves that, whenever

$$Kx = Ky, \tag{3.24}$$

then $x\theta = y\theta$. It is only under these circumstances that ϕ is well-defined by (3.23) and disastrous inconsistencies are prevented. Now (3.24) is equivalent to the statement that $y = ux$, where $u \in K$. By the definition of K, $u\theta = 1'$ and therefore $y\theta = u\theta\, x\theta = x\theta$, as required. We can now proceed to show that ϕ has all the properties we seek.

(1) ϕ is a homomorphism; for,

$$(Kx)\phi(Ky)\phi = x\theta y\theta = (xy)\theta = (Kxy)\phi.$$

(2) ϕ is surjective; this is obvious, because in (3.23) x can be any element of G, so that all image elements $x\theta$ are reached by ϕ.

(3) ϕ is injective; we have to prove that

$$(Kx)\phi = (Ky)\phi \tag{3.25}$$

implies that $Kx = Ky$. If (3.25) is assumed, then by the definition of ϕ, $x\theta = y\theta$. This means that $xy^{-1} \in K$, which is equivalent to $Kx = Ky$.

---

* As we shall see, there are several isomorphism theorems. Unfortunately, there is no unanimity in the literature about their numbering.

This completes the proof of the theorem, which may be para-phrased by the statement that every homomorphic image of G is isomorphic with a quotient group of G, namely the quotient by the kernel.

To round off the picture, we point out that any normal subgroup of G occurs as the kernel of a suitable homomorphism. Let $N \lhd G$ and consider the map $v: G \to G/N$ defined by

$$xv = Nx \quad (x \in G). \tag{3.26}$$

Thus, in this case, $G' = G/N$. It is easy to verify that (3.26) is a homomorphism; for

$$(xv)(yv) = NxNy = Nxy = (xy)v.$$

Clearly, v is in fact an epimorphism, because in (3.26) x can be any element of G, and so the whole of G/N is covered. The kernel of v consists of those elements $u \in G$ for which $Nu = N$ (the identity in G/N); this is equivalent to the condition that $u \in N$. Thus ker $v = N$. The map defined in (3.26) is called the **natural map** of G onto G/N.

By way of illustration, let us return to example (iii) on p. 65, where $G = GL(n, F)$. If $a \in G$, we consider the homomorphism

$$\delta: G \to F$$

defined by

$$a\delta = \det a.$$

in this case, $G\delta = F^\times$, and the kernel consists of the group $U = \{a | \det a = 1\}$. This is automatically a normal subgroup of G. By Theorem 8, we have that

$$G/U \cong F^\times,$$

as before.

The First Isomorphism Theorem gives us a clearer insight into the effect of the homomorphism $\theta: G \to G'$: all the elements of Kx have the image $x' = x\theta$; in particular, when $|K|$ is finite, the image group $G\theta$ is covered exactly $|K|$ times. Moreover, when the index $[G:K]$ is finite, we have that

$$|G\theta| = [G:K]. \tag{3.27}$$

22. Subgroups of Quotient Groups. Let $N \lhd G$ be a normal subgroup of G. We wish to examine the subgroups of G/N and study their relationship with subgroups of G. In order to avoid confusion it will be necessary temporarily to introduce a slightly more elaborate notation for the elements of G/N. A typical element of G/N will now be written as (Nx) to distinguish it from the subset Nx consisting of $|N|$ elements of G. A subgroup A' of G/N is a collection of elements, say

$$A' = (N) \cup (Na) \cup (Nb) \cup \ldots, \qquad (3.28)$$

which satisfy the group axioms relative to the composition law in G/N. Removing the brackets we obtain a subset

$$A = N \cup Na \cup Nb \cup \ldots \qquad (3.29)$$

of G. We claim that A is, in fact, a subgroup of G. Evidently $N \subset A$ and hence $1 \in A$. Next, if x and y are elements of A, then (Nx) and (Ny) are elements of A'; since A' is a group, $(Nxy) \in A'$, which in turn implies that $xy \in A$. Finally, if $x \in A$, then $(Nx^{-1}) \in A'$, and hence $x^{-1} \in A$. Thus we have demonstrated that A is a group, and more precisely that

$$N \leqq A \leqq G. \qquad (3.30)$$

Conversely, if A is any subgroup of G satisfying (3.30), we note that in fact $N \lhd A$; for the relation $x^{-1}Nx = N$ holds for all elements x of G, and in particular for all those x which lie in A. It is therefore legitimate to form the quotient group A/N. Now if (3.29) is the coset decomposition of A with respect to N, then, by inserting brackets, we obtain A/N, which is a subgroup of G/N. Clearly, distinct subgroups A' and B' of G/N give rise to distinct subgroups A and B of G, each containing N, and conversely. Thus there is a one-to-one correspondence between the subgroups of G/N and those subgroups of G which contain N.

It is of interest to enquire how the normal subgroups of G/N can be described in this context. We may assume that such a subgroup is expressed in the form A/N, where A satisfies (3.30). Now

$$A/N \lhd G/N \qquad (3.31)$$

if and only if, for every $x \in G$ and for every $a \in A$,

$$(Nx)^{-1}(Na)(Nx) = (Nx^{-1}ax) \in A/N,$$

and this is equivalent to the condition that

$$x^{-1}ax \in A,$$

in other words, that $A \lhd G$. We summarize these results as follows.

PROPOSITION 10. *All subgroups of G/N can be expressed as A/N, where*

$$N \leqq A \leqq G,$$

and $A/N \lhd G/N$ if and only if

$$N \lhd A \lhd G.$$

Moving within G/N and assuming that (3.31) is satisfied, we can construct the quotient group

$$(G/N)/(A/N).$$

Fortunately, the complication engendered by forming a quotient of quotient groups is rendered less formidable by the next theorem.

THEOREM 9. **(Second Isomorphism Theorem).** *Let $N \lhd G$ and suppose that A is a normal subgroup of G such that*

$$N \lhd A \lhd G.$$

Then

$$(G/N)/(A/N) \cong G/A. \tag{3.32}$$

Proof. Consider the map

$$\phi: G/N \to G/A,$$

which is defined by the rule that

$$(Nx)\phi = (Ax) \quad (x \in G). \tag{3.33}$$

First, we must check that (3.33) is indeed a meaningful definition. The element x on the left may be replaced by ux, where $u \in N$, without altering the coset Nx, and we have to show that this substitution does not alter the right-hand side of (3.33). Since $N \leqq A$, we have that $u \in A$, whence $Au = A$ (Proposition 3, p. 31) and consequently $Aux = Ax$, as required. Next, we observe that ϕ is a homomorphism; for

$$(Nx)\phi(Ny)\phi = (Ax)(Ay) = (Axy) = (Nxy)\phi,$$

by virtue of the normality of A. Clearly, ϕ is surjective, because, in (3.33), x is an arbitrary element of G, and so all cosets of A will appear on the right of (3.33); thus

$$(G/N)\phi = G/A. \tag{3.34}$$

It remains to find ker ϕ. Now $(Nx) \in \ker \phi$ if and only if $(Ax) = (A)$, the identity element of G/A. This is equivalent to the condition that $x \in A$. Hence ker ϕ is the union of the cosets (Na), where a ranges over A, in other words,

$$\ker \phi = A/N. \tag{3.35}$$

Using (3.34) and (3.35), we see that (3.32) is an immediate consequence of the First Isomorphism Theorem.

We return once more to the general case of a homomorphism

$$\theta: G \to G'. \tag{3.36}$$

and ask how this map affects a given subgroup A of G. This means that we consider the restriction map

$$\theta_A: A \to G' \tag{3.37}$$

defined by the obvious rule that

$$a\theta_A = a\theta \quad (a \in A).$$

It may seem rather pedantic to press a new symbol, θ_A, into service, and indeed the difference between θ and θ_A is sometimes ignored. Yet it may be insisted that (3.36) and (3.37) are distinct maps because they have different 'domains of definition'. As in all homomorphisms, the image group

$$A' = A\theta_A \quad (= A\theta)$$

is a subgroup of G', whilst the kernel clearly consists of those elements of A which lie in the kernel of θ, that is

$$\ker \theta_A = A \cap \ker \theta. \tag{3.38}$$

It is worthwhile considering in more detail what happens when the natural epimorphism

$$v: G \to G/N, \quad xv = (Nx),$$

is restricted to a subgroup A of G. The image group can be written explicitly as

$$A' = Av_A = \bigcup_a (Na), \tag{3.39}$$

where a ranges over A, although it should be noted that the union may contain redundant terms. On the other hand, A' is a subgroup of G/N and, as we have seen on p. 70, must be of the form $A' = B/N$, where $N \leq B \leq G$. In the present situation we cannot say that $B = A$, because A need not contain N so that A/N would be meaningless. The rule for finding B was given on p. 70 and consists in removing the brackets in (3.39), thus

$$B = \bigcup Na, \quad (a \in A).$$

This may be more concisely expressed in the notation of subsets, namely

$$B = NA.$$

It is instructive to verify by a different method that B is a subgroup. For since N is normal, $Na = aN$ for each $a \in A$ and therefore $NA = AN$. Hence by the Product Theorem (p. 52), B is a group. Thus we note that

$$Av_A = NA/N. \tag{3.40}$$

Next, since ker $v = N$, we deduce from (3.38) that

$$\ker v_A = A \cap N, \tag{3.41}$$

and we note that, being a kernel, $A \cap N$ is normal in A.

The First Isomorphism Theorem, applied to v_A, states that

$$A/\ker v_A \cong Av_A.$$

Substituting from (3.40) and (3.41) we recast the result as follows.

THEOREM 10. (**Third Isomorphism Theorem**). *Let N be a normal subgroup and A an arbitrary subgroup of G. Then*

$$\frac{A}{A \cap N} \cong \frac{NA.}{A}$$

It is worthwhile considering the interior direct product (p. 42) in the context of normal subgroups. If

$$G = H \times K, \tag{3.42}$$

then each element of K commutes with each element of H; thus if $v \in K$ we certainly have that $v^{-1}Hv = H$. Also if $u \in H$, then $u^{-1}Hu = H$ (Proposition 3, p. 31). Since any element $x \in G$ can be expressed as $x = uv$, it follows that $x^{-1}Hx = H$. Therefore $H \lhd G$ and, similarly, $K \lhd G$, that is, *in a direct product, each factor is a normal subgroup*.

Next, we observe that

$$G/K \cong H. \tag{3.43}$$

This follows at once from the Third Isomorphism Theorem, if we put $K = N$, $H = A$ and note that $KH = H \times K = G$, $H \cap K = \{1\}$. Alternatively, using a more direct argument we remark that any coset of K in G is of the form Ku, where $u \in H$. For if $x = uv$ ($u \in H$, $v \in K$) is an arbitrary element of G, then $Kx = Kuv = Kvu = Ku$, because $Kv = K$. Also, if $Ku_1 = Ku_2$, where $u_1, u_2 \in H$, then $u_1 u_2^{-1} \in H \cap K = \{1\}$ and therefore $u_1 = u_2$. Hence

$$Ku \to u$$

furnishes a bijective homomorphism between the groups G/K and H.
  Obviously,

$$H \times K \cong K \times H$$

by virtue of the correspondence

$$(u, v) \leftrightarrow (v, u), \ (u \in H, v \in K).$$

23. The Derived Group. Corresponding to any two elements x and y of a group G, we define their **commutator** as

$$[x, y] = x^{-1}y^{-1}xy.$$

Clearly, $[x, y] = 1$ if and only if $xy = yx$. We are interested in the set of all commutators, as x and y range over G. This set, however, does not in general form a group, because the product of two commutators cannot always be expressed as a single commutator. (It is a curious fact that this failure becomes apparent only in fairly complicated groups.) In any event, we can construct the group which is generated by all the commutators; this group is called the **derived group** or **commutator group** of G and is usually denoted by G', that is

$$G' = \text{gp} \{[x, y] \big| x, y \in G\}. \tag{3.44}$$

Thus a typical element of G' is a finite product of commutators. Evidently $G' = \{1\}$ if and only if G is Abelian. The main properties of G' are collected in the following theorem:

THEOREM 11. (i) *The derived group G' is a normal subgroup of G, and G/G' is Abelian.* (ii) *If H is any normal subgroup of G such that G/H is Abelian, then $G' \leq H$.*

  Proof. (i) In order to show that $G' \lhd G$, it suffices to prove that, for all $t \in G$,

$$[x, y]^t \in G'$$

(see (3.2) and (3.10)). By virtue of the rules (3.3) and (3.3)′ we have that

$$[x, y]^t = [x^t, y^t].$$

Since the right-hand side is a commutator, it belongs to G'. Hence $G' \lhd G$. Next, we shall prove that the cosets $G'x$ and $G'y$ commute or, in a different notation, that

$$[G'x, G'y] = G'.$$

Now

$$[G'x, G'y] = (G'x)^{-1}(G'y)^{-1}(G'x)(G'y)$$
$$= G'x^{-1}y^{-1}xy = G'[x, y] = G',$$

because $[x, y] \in G'$. Thus G/G' is Abelian.

(ii) If $H \lhd G$, we can repeat the above computation with H in place of G' and find that

$$[Hx, Hy] = H[x, y].$$

When G/H is Abelian, the left-hand side reduces to the unit element of G/H, that is to H, and we deduce that $[x, y] \in H$. Since x and y are arbitrary, it follows that each generator of G' lies in H, whence $G' \leqq H$.

We conclude this section by proving the following result.

PROPOSITION 11. *Let A and B be normal subgroups of G such that $A \cap B = \{1\}$. Then each element of A commutes with each element of B.*

Proof. Consider the commutator

$$c = a^{-1}b^{-1}ab,$$

where a and b are arbitrary elements of A and B respectively. Since $A \lhd G, a_1 = b^{-1}ab \in A$ and therefore $c = a^{-1}a_1 \in A$; similarly $c \in B$. Thus $c \in A \cap B = \{1\}$, whence $c = 1$, that is $ab = ba$.

24. Automorphisms. An interesting type of isomorphism of G occurs when the image group coincides with G. An isomorphism

$$\alpha: G \to G$$

of G onto itself is called an **automorphism** of G. In particular α is a bijective map of G onto itself, that is α permutes the elements of G.

Of course, the converse is not true, because α must, in addition, satisfy the relation

$$(xy)\alpha = (x\alpha)(y\alpha) \quad (x, y \in G). \tag{3.45}$$

Applying the observations of p. 18, we conclude that the collection of all automorphisms of G forms a group under composition of maps. If

$$\beta: G \to G$$

is another automorphism, we denote the product of α and β by $\alpha\beta$ instead of $\alpha \circ \beta$. Thus the effect of $\alpha\beta$ on an element $x \in G$ is defined by the rule

$$x(\alpha\beta) = (x\alpha)\beta.$$

The group of all automorphisms of G is denoted by $A(G)$ and is called the **automorphism group** of G. The unit element of G is the identity automorphism ι which leaves every element of G fixed, that is

$$x\iota = x \quad (x \in G). \tag{3.46}$$

The inverse of α is denoted by α^{-1}. Thus $x\alpha^{-1}$ is the unique element y of G which satisfies $y\alpha = x$; such an element exists for each x, because α is surjective.

Since α is injective, $\ker \alpha = \{1\}$. This implies that α preserves the order of each element. For if $y = x\alpha$ and $x^m = 1$, then by (3.45)

$$1 = x^m\alpha = (x\alpha)^m = y^m.$$

Hence the order of y is not less than the order of x. Using α^{-1} instead of α we establish the opposite inequality. Hence x and y have the same order, which may be infinite. With a fixed element t of G we associate the map

$$\tau: G \to G$$

given by

$$x\tau = x^t(= t^{-1}xt) \quad (x \in G). \tag{3.47}$$

Equation (3.3) shows that τ is a homomorphism of G into itself; it is, in fact, an automorphism. For, $x^t = 1$ implies that $x = 1$; thus the kernel of τ reduces to the unit element whence, by Proposition 9 (p. 67), τ is injective. Again, if y is any element of G, there exists an x such that $x^t = y$, namely $x = tyt^{-1}$; hence τ is surjective. An automorphism like (3.47), which is induced by conjugation, is called an **inner automorphism** of G. An automorphism which is not an inner automorphism, is called an **outer automorphism**.

Next, we show that the collection $I(G)$, of all inner automorphisms forms a group under composition of maps. Thus, let σ be another inner automorphism, given by

$$x\sigma = s^{-1}xs \quad (x \in G).$$

Then

$$x\tau\sigma = (t^{-1}xt)\sigma = s^{-1}t^{-1}xts$$

$$= (ts)^{-1}x(ts),$$

that is

$$x^t x^s = x^{ts}. \tag{3.48}$$

Hence the composite map $\tau\sigma$ corresponds to conjugation by ts. This proves the closure of $I(G)$. Evidently, $\iota \in I(G)$, as we may take $t = 1$, and τ^{-1} corresponds to conjugation by t^{-1}, that is

$$x\tau^{-1} = txt^{-1} \quad (x \in G).$$

More precise information about the group $I(G)$ is furnished by the following proposition.

PROPOSITION 12. *Let Z be the centre of G. Then*

$$I(G) \cong G/Z.$$

Proof. The correspondence between an element t and the inner automorphism τ, which t induces, is formalized by the map

$$\Phi: G \to I(G), \tag{3.49}$$

defined by

$$t\Phi = \tau \quad (t \in G).$$

Equation (3.48) now states that $(ts)\Phi = (t\Phi)(s\Phi)$, that is Φ is a homomorphism. It is obvious that Φ is surjective, because every inner automorphism is obtained by operating with Φ on a suitable element of G; thus

$$G\Phi = I(G).$$

Next, we wish to find ker Φ. Now $t \in$ ker Φ if and only if the inner automorphism induced by t is the identity automorphism

$$x^t = x \quad (x \in G).$$

But this equation is equivalent to the statement that $t \in Z$. Thus $\ker \Phi = Z$. An application of the First Isomorphism Theorem immediately proves the assertion.

In an Abelian group all inner automorphisms collapse into the identity map, the outer automorphisms being then the only possible non-trivial ones. The following simple examples will serve as illustrations:

(1) *The infinite cyclic group* $C = \text{gp}\{x\}$. Any automorphism α is determined as soon as $x\alpha$ is known, say $x\alpha = x^s$, where s is an integer. If x^k is an arbitrary element of C, then $x^k\alpha = (x\alpha)^k = x^{ks}$. Thus the image group is $C\alpha = \text{gp}\{x^s\}$. But, for an automorphism, $C\alpha = C$. Hence we must have that $s = 1$ or $s = -1$. Both cases are possible, the first being the identity map. Thus C possesses precisely two automorphisms.

(2) *The finite cyclic group* $C_m = \text{gp}\{x; x^m = 1\}$. As before, only $x\alpha = x^s$ need be specified. Under any automorphism, the order of an element is preserved. Hence x^s must be of order m. This happens if and only if $(s, m) = 1$ (see Proposition 2, p. 17), and any such choice of s gives rise to an automorphism. Thus C_m has $\phi(m)$ automorphisms, where ϕ is the Euler function defined on p. 10.

(3) *The four-group* $V = \text{gp}\{a, b; a^2 = b^2 = 1, ab = ba\}$. This group has three elements of order two, which can only be permuted under α. It turns out that each of the six permutations determines an automorphism; for if the three elements of order two are denoted by x, y, z (in any arrangement), then $xy = z$. So if $x\alpha = x'$, $y\alpha = y'$, $z\alpha = z'$, we shall have that $x'y' = z'$. It follows that V has six automorphisms and that $A(V) \cong S_3$ (see p. 22).

If α is an automorphism of G, we can study its effect on a subgroup H of G. In all cases the image of H under α is a subgroup $H\alpha$ of G. If

$$H\alpha = H \tag{3.50}$$

holds (as an equation between subsets), then we say that H is invariant under α. For example, H is normal in G if and only if it is invariant under all inner automorphisms. In that case, for every $t \in G$, the map $H \to t^{-1}Ht$ is an automorphism of H.

A subgroup H which is invariant under all automorphisms is called a **characteristic subgroup**. Of course, all characteristic subgroups are normal. For example, the centre, Z, is a characteristic subgroup; for if $t \in Z$ then $tx = xt$ holds for all $x \in G$. Hence, for any $\alpha \in A(G)$, $(t\alpha)(x\alpha) = (x\alpha)(t\alpha)$; but since α is surjective, $x\alpha$ can be made equal to any element $y \in G$. Thus $(t\alpha)y = y(t\alpha)$ holds for all $y \in G$, that is

$$Z\alpha \subset Z.$$

On replacing α by α^{-1} we arrive at the opposite inequality, so that $Z\alpha = Z$.

We conclude this section by proving the following proposition:

PROPOSITION 13. *Suppose that N is a normal subgroup of G and that H is a characteristic subgroup of N. Then H is normal in G.*

Proof. Let $t \in G$. Then, as we have just remarked, the map τ defined in (3.47) is an automorphism of N. Therefore, since H is characteristic in N, we have that $H\tau = H$. Thus $t^{-1}Ht = H$, that is H is normal in G.

Exercises

(1) Show that conjugate elements have the same order.

(2) Two classes (a) and (a^{-1}), which are generated by inverse elements, are called inverse classes. Prove that (i) inverse classes contain the same number of elements and (ii) a group of even order includes at least one class, besides that consisting of the unit element, which is identical with its inverse class.

(3) Let $G = GL(n, F)$ (see p. 8), where $n \geq 2$ and F is an infinite field. Prove that the centre of G consists of all scalar multiples of the unit matrix.

(4) Find the centre, Z, of the dihedral group of order 8 (Table (xi), p. 51) and determine the structure of G/Z.

(5) Show that the set $T = \{t = (t_{ij}) | t_{ij} = 0 \text{ if } i > j,\ t_{ii} \neq 0\}$ of $n \times n$ non-singular upper-triangle matrices over a field forms a group under matrix multiplication. Prove that the subset E for which $t_{ii} = 1$ $(i = 1, 2, \ldots, n)$, is a normal subgroup of T, and show that $T/E \simeq D$, where D is the set of nonsingular diagonal matrices.

(6) Prove that if H is a subgroup of G, then the number of subgroups conjugate with H is equal to $[G:N(H)]$.

(7) Let N be a normal subgroup of G of finite index n. Given an element $t \in G$, let h be the least positive integer such that $t^h \in N$. Prove that $h | n$; also show that, if t is of finite order r, then $h | r$.

(8) Suppose that a and b are elements of a group such that their commutator $c = [a, b]$ commutes with both a and b. Show that, if k is a positive integer, (i) $a^k b = ba^k c^k$ and (ii) $(ab)^k = b^k a^k c^{\frac{1}{2}k(k+1)}$.

(9) Let N be a normal subgroup of G of finite index n. Show that, if A is any subgroup of G, then $s = [A:A \cap N]$ is finite and $s|n$.

(10) Find the derived group of (i) the dihedral group of order 8 and (ii) the quaternion group.

(11) Prove that the centralizer of a normal subgroup of G is a normal subgroup of G.

(12) Show that in an Abelian group the map $x\theta = x^{-1}$ is an automorphism.

(13) Prove that $I(G)$ is a normal subgroup of $A(G)$.

(14) Show that G' is a characteristic subgroup of G.

IV. Finitely Generated Abelian Groups

25. Preliminaries. In this chapter we shall be concerned only with Abelian groups, and it will be convenient to employ the additive notation (see pages 6 to 7). We recall that, in this case, all subgroups are normal; if $H \leq G$, the quotient group G/H consists of the cosets $H+x$ ($x \in G$). The group G is said to be **finitely generated** abbreviated as f.g., if there exist finitely many elements $u_1, u_2, \ldots, u_n$ in G, called **generators**, such that

$$G = \text{gp} \{u_1, u_2, \ldots, u_n\}.$$

Then every element x of G is a finite sum of some of these generators or their negatives (inverses) in any order, repetitions being permitted. However, by virtue of the commutative law, we can collect terms which involve the same generator, and we can write

$$x = a_1 u_1 + a_2 u_2 + \ldots + a_n u_n, \tag{4.1}$$

where the a_i are integers (positive, negative or zero). Conversely, for any choice of integral coefficients, (4.1) represents an element of G. But the generators are not assumed to be irredundant, and even when they are, they may satisfy non-trivial relations

$$c_1 u_1 + c_2 u_2 + \ldots + c_n u_n = 0, \tag{4.2}$$

in which not all coefficients are zero. Since fractional coefficients are disallowed, we cannot in general 'solve' (4.2) for one of the u's in terms of the others.

In the sequel we shall frequently have occasion to modify a given set of generators, and it is therefore of interest to study the conditions under which two sets of elements can serve as generators of the same Abelian group. Thus suppose that

$$G = \text{gp} \{u_1, u_2, \ldots, u_n\} = \text{gp} \{v_1, v_2, \ldots, v_m\}. \tag{4.3}$$

In order that (4.3) should hold it is necessary and sufficient that each u is expressible in terms of the v's and that, conversely, each v is expressible in terms of the u's. Thus we have equations of the form

$$\left. \begin{array}{ll} u_i = \sum_{j=1}^{m} p_{ij}v_j & (i = 1, 2, \ldots, n) \\[2mm] v_j = \sum_{k=1}^{n} q_{jk}u_k & (j = 1, 2, \ldots, m) \end{array} \right\}, \qquad (4.4)$$

where the matrices $\mathbf{p} = (p_{ij})$ and $\mathbf{q} = (q_{jk})$ have integral coefficients or, more briefly, are *integral matrices*. We refer to the system of equations (4.4) as a *transformation* of the set of generators $u_1, u_2, \ldots, u_n$ into the set $v_1, v_2, \ldots, v_n$.

The following types of transformation of generators are most common.

(α) The generators may be permuted in any manner.

(β) If $i \neq j$, the generator u_i may be replaced by $u_i + hu_j$, where h is an arbitrary integer, all other generators remaining unaltered.

(γ) Any generator u_i may be replaced by $-u_i$,

(δ) If a generator is zero, it may be omitted.

The operations (α), (β) and (γ) are called elementary transformations. Let us check that (β) does indeed satisfy (4.4). For simplicity, assume that $i = 1, j = 2$ so that we have the transformation

$$v_1 = u_1 + hu_2, v_2 = u_2, \ldots, v_n = u_n,$$

which is inverted by the equations

$$u_1 = v_1 - hv_2, u_2 = v_2, \ldots, u_n = v_n.$$

The operations listed above can be applied repeatedly until we obtain a system of generators which is convenient for our purpose.

A minor technical point is worth mentioning here. In order to prove that a subset X of G forms a subgroup it is sufficient to verify that whenever x and y belong to X, then

$$x - y \in X.$$

For if this is true, we may take $x = y$ and find that $0 \in X$; again choosing $x = 0$ we find that $-y \in X$; finally on replacing y by $-y$ we deduce that $x + y \in X$, thus verifying all the conditions (see §9, p. 30) for X to be a subgroup of G.

We shall confine ourselves to finitely generated Abelian groups, and it is our aim to give a complete description of all possible types of groups in this class (up to isomorphisms). This will be accomplished by breaking G up into a **direct sum** of certain subgroups, in analogy to the notion of a direct product (§13, p. 41). A direct sum is written as

$$G = H \oplus K. \tag{4.5}$$

We are here concerned with internal direct sums. Thus (4.5) means that there are subgroups H and K of G with the following properties: the elements of G consist of all possible sums

$$x = u + v, \tag{4.6}$$

where u and v independently range over H and K respectively, and this representation is unique. Thus if

$$u_1 + v_1 = u_2 + v_2, \tag{4.7}$$

where $u_1, u_2 \in H$ and $v_1, v_2, \in K$, then $u_1 = u_2$ and $v_1 = v_2$. In particular, if $u_0 + v_0 = 0$, where $u_0 \in H$ and $v_0 \in K$, then $u_0 = v_0 = 0$. Conversely, this statement ensures the uniqueness of (4.6); for (4.7) implies that $(u_1 - u_2) + (v_1 - v_2) = 0$ and hence $u_1 = u_2$, $v_1 = v_2$. Again, in order to prove (4.5) it is sufficient to show that

(i) $G = H + K$ and   (ii) $H \cap K = \{0\}$.

The second condition is certainly satisfied when H and K are finite groups of coprime orders.

When G is expressed as the direct sum of several subgroups, we use the notation

$$G = \sum_{i=1}^{n} \oplus H_i = H_1 \oplus H_2 \oplus \ldots \oplus H_r, \tag{4.8}$$

and we recall that, up to isomorphism, the formation of direct sums is both commutative and associative. Indeed, (4.8) states that G is isomorphic with the group whose elements are all r-tuples $(u_1, u_2, \ldots, u_r)$, where u_i ranges over H_i and composition is carried out separately in each component.

For example, (4.8) will certainly hold if
(i) $G = H_1 + H_2 + \ldots + H_r$
and
(ii) the orders of H_i and H_j $(i \neq j)$ are coprime.
For in that case it is obvious that

$$H_i \cap H_1 + \ldots + H_{i-1} + H_{i+1} + \ldots + H_r = \{0\}.$$

26. Finitely generated Free Abelian Groups. In this section we study f.g. Abelian groups.

$$F = \text{gp } \{u_1, u_2, \ldots, u_n\}, \tag{4.9}$$

in which the generators satisfy no non-trivial relations, that is we make the hypothesis that

$$c_1 u_2 + c_2 u_2 + \ldots + c_n u_n = 0 \tag{4.10}$$

always implies that $c_1 = c_2 = \ldots = c_n = 0$. If such a system of generators exists, we call F a **free Abelian group**. More precisely we say that F is **freely generated** by $u_1, u_2, \ldots, u_n$. Such a system of generators is then called a set of free generators, and we use the notation

$$F = \langle u_1, u_2, \ldots, u_n \rangle. \tag{4.11}$$

Thus (4.11) is equivalent to the statement that the elements of F are *uniquely* expressible in the form

$$x = a_1 u_1 + a_2 u_2 + \ldots + a_n u_n, \text{ where the } a_i \text{ are arbitrary integers.} \tag{4.12}$$

It is clear that in a free Abelian group all elements, other than zero, are of infinite order. For if $x \neq 0$ and $h > 0$, the equation $hx = 0$ would immediately lead to a non-trivial relation for the generators. In particular, each generator is of infinite order, and (4.11) is equivalent to

$$F = \text{gp } \{u_1\} \oplus \text{gp } \{u_2\} \oplus \ldots \oplus \text{gp } \{u_n\}, \tag{4.13}$$

a direct sum of n infinite cyclic groups.

It is easy to give an example of a free Abelian group with n free generators: let Z^n be the collection of all n-tuples $x = [a_1, a_2, \ldots, a_n]$, where $a_1, a_2, \ldots, a_n$ independently range over all integers. Define the law of composition in Z^n as componentwise addition, thus making Z^n an Abelian group. The special n-tuples

$$u_1 = [1, 0, \ldots, 0], u_2 = [0, 1, \ldots, 0], \ldots, u_n = [0, 0, \ldots, 1]$$

generate Z^n because, for any $x \in Z^n$,

$$x = a_1 u_1 + a_2 u_2 + \ldots + a_n u_n.$$

Moreover, these generators are free because

$$c_1 u_1 + c_2 u_2 + \ldots + c_n u_n = [c_1, c_2, \ldots, c_n] = 0,$$

implies that $c_1 = c_2 = \ldots = c_n = 0$.

We will now examine the connection between different sets of free generators. If

$$F = \langle u_1, u_2, \ldots, u_n \rangle = \langle v_1, v_2, \ldots, v_m \rangle, \qquad (4.14)$$

the two systems of generators are linked by equations (4.4). But since the generators are free, we have more precise information at our disposal. Eliminating the v_j in (4.4) we find that

$$u_i = \sum_{j=1}^{m} \sum_{k=1}^{n} p_{ij} q_{jk} u_k, \quad (i = 1, 2, \ldots, n).$$

This would be a non-trivial relation between the u's, unless corresponding coefficients agree on both sides. Thus we must have that

$$\sum_{j=1}^{m} p_{ij} q_{jk} = \delta_{ik} \quad (i, k = 1, 2, \ldots, n),$$

where $\delta_{ik} = 0$ if $i \neq k$ and $\delta_{ii} = 1$; or in matrix notation

$$pq = i_n, \qquad (4.15)$$

where i_n is the unit matrix of degree n. Similarly, by eliminating the u's,

$$qp = i_m. \qquad (4.16)$$

The reader who has some knowledge of Linear Algebra, will have no difficulty in deducing from (4.15) and (4.16) that $m = n$; alternatively, we can verify this fact by computing the sum of the diagonal elements in (4.15) and (4.16), namely

$$\sum_{i=1}^{n} \sum_{j=1}^{m} p_{ij} q_{ji} = n, \quad \sum_{j=1}^{m} \sum_{i=1}^{n} q_{ji} p_{ij} = m.$$

Since the expressions on the left-hand sides of these equations are equal it follows that $m = n$. Thus the number of free generators is an invariant for F, that is, it is the same for every system of free generators. This number is called the **rank** of F. Moreover, two f.g. free Abelian groups are isomorphic if and only if they have the same rank; for, if the rank is n, either group is then isomorphic with the group displayed in (4.13), or, alternatively, with the group of all integral n-tuples.

Taking determinants in (4.15) or (4.16) we find that

$$(\det p)(\det q) = 1. \qquad (4.17)$$

But the coefficients of p and q are integers, and their determinants are therefore also integers. Hence we deduce from (4.17) that $\det p = \det q = \pm 1$, that is p and q are unimodular matrices and therefore possess integral inverses (see ex. (iv), (c) p. 9). Thus the transition from one set of free generators to another is accomplished by a unimodular transformation

$$u_i = \sum_{j=1}^{n} p_{ij} v_j \quad (i = 1, 2, \ldots, n), \tag{4.18}$$

and it is clear that any unimodular matrix p can be used for this purpose. For (4.18) can be inverted by equations

$$v_j = \sum_{k=1}^{n} q_{jk} u_k \quad (j = 1, 2, \ldots, n) \tag{4.19}$$

where $q = p^{-1}$ is again an integral matrix, thus verifying (4.4).

The operations (α), (β) and (γ) described on p. 82 are simple instances of unimodular transformations. When several of these operations are carried out in succession, the corresponding matrices are multiplied.

The highest common factor (HCF) of a set of integers $a_1, a_2, \ldots, a_n$, not all of which are zero, is written as

$$(a_1, a_2, \ldots, a_n)$$

and is, by definition, a positive integer. In particular, when $(a_1, a_2, \ldots, a_n) = 1$, we say that the integers are coprime. It is clear that in a unimodular matrix the coefficients which form a row or a column, must be coprime. For if the determinant of the matrix is expanded with respect to a row (column), it is plain that the determinant is divisible by the HCF of this row (column). However, by hypothesis, the determinant equals ± 1, whence the HCF can only be equal to one. Thus if a new set of free generators is introduced by (4.19), each new generator is a linear combination of the old generators with coprime coefficients. The following proposition establishes a partial converse of this fact.

PROPOSITION* 14. *Let* $F = \langle u_1, u_2, \ldots, u_n \rangle$ *and suppose that* $v = b_1 u_1 + b_2 u_2 + \ldots + b_n u_n$ *is an element of* F *such that*

$$(b_1, b_2, \ldots, b_n) = 1. \tag{4.20}$$

---

* See R. Rado, 'A proof of the basis theorem for finitely generated Abelian groups', *Journal of the London Mathematical Society* (1951), **26**, 74–75.

Then there exist elements $v_2, v_3, \ldots, v_n$ of F such that

$$F = \langle v, v_2, v_3, \ldots, v_n \rangle. \qquad (4.21)$$

In other words, (4.20) is the necessary and sufficient condition that an element can be incorporated in a set of free generators.

Proof. Let $s = |b_1| + |b_2| + \ldots + |b_n|$. If $s = 1$, then $v = \pm u_j$, for some j, and it is obvious that v can be included in a set of free generators. We now use induction on s, at the same time reserving the right to change the generators of F until (4.21) is established. If $s > 1$, at least two b's are non-zero, since otherwise $(b_1, b_2, \ldots, b_n) > 1$. Without loss of generality it may be assumed that $b_1 \geq b_2 > 0$, because this condition can always be satisfied by permuting the generators and changing their signs (operations (α) and (γ), p. 82). Now let

$$u_1' = u_1, u_2' = u_2 + u_1, u_j' = u_j \quad (j \geq 3).$$

It is clear that $F = \langle u_1', u_2', \ldots, u_n' \rangle$ (operation (β)). The expression for v now becomes $v = (b_1 - b_2)u_1' + b_2 u_2' + \ldots + b_n u_n'$. Evidently, $(b_1 - b_2, b_2, b_3, \ldots, b_n) = 1$, but

$$|b_1 - b_2| + |b_2| + |b_3| + \ldots + |b_n| < s.$$

Hence by the inductive hypothesis, v can be included in a set of free generators.

We now turn our attention to subgroups H of an f.g. free Abelian group F. It may be asked whether H is also f.g. and free. The question is answered in the affirmative by the next theorem, which is crucial for the theory of Abelian groups; it also establishes the deeper result that the generators of H can be expressed in a surprisingly simple manner provided that the generators of F are chosen conveniently.

THEOREM 12. *Let F be an f.g. free Abelian group of rank n, and let H be a non-zero subgroup of F. Then H is a f.g. free Abelian group of rank $m \leq n$. It is possible to choose a set of free generators $v_1, v_2, \ldots, v_n$ for F in such a way that*

$$H = \langle h_1 v_1, h_2 v_2, \ldots, h_m v_m \rangle, \qquad (4.22)$$

where $h_1, h_2, \ldots, h_m$ are positive integers satisfying the relations $h_i | h_{i+1} \ (i = 1, 2, \ldots, m-1)$.

Proof. (i) Suppose that F is originally given in terms of the free generators $u_1, u_2, \ldots, u_n$. With every non-zero element $x = a_1 u_1 + a_2 u_2 + \ldots + a_n u_n$ of F we associate the HCF of its coefficients relative to this set of generators, say

$$\delta(x) = (a_1, a_2, \ldots, a_n).$$

This number is, however, independent of the choice of generators. For if $u_1', u_2', \ldots, u_n'$ is another set of free generators of F we have that $u_i = \Sigma_j p_{ij} u_j'$, where (p_{ij}) is a unimodular matrix. Then $x = a_1' u_1' + a_2' u_2' + \ldots + a_n' u_n'$, where $a_j' = \Sigma_i a_i p_{ij}$. Hence any common factor of the a_i must divide all the a_j', whence

$$(a_1', a_2', \ldots, a_n') \geqq (a_1, a_2, \ldots, a_n).$$

If we interchange the roles of the two sets of generators by inverting the matrix (p_{ij}), we can establish the opposite inequality. Hence $(a_1', a_2', \ldots, a_n') = (a_1, a_2, \ldots, a_n)$, which proves the invariance of $\delta(x)$.

(ii) Among the non-zero elements of H let

$$y_1 = b_1 u_1 + b_2 u_2 + \ldots + b_n u_n$$

be such that δ attains its least value, say $\delta(y_1) = h_1 \geqq 1$. We can then write $y_1 = h_1(c_1 u_1 + c_2 u_2 + \ldots + c_n u_n) = h_1 v_1$, where $v_1 = c_1 u_1 + c_2 u_2 + \ldots + c_n u_n$ is an element of F with the property that $(c_1, c_2, \ldots, c_n) = 1$. By Proposition 14 there exist elements $v_2', v_3', \ldots, v_n'$ such that

$$F = \langle v_1, v_2', v_3', \ldots, v_n' \rangle. \tag{4.23}$$

Using this set of generators let $y = d_1 v_1 + d_2 v_2' + \ldots + d_n v_n'$ be an arbitrary element of H. We know that

$$y_1 = h_1 v_1 \in H, \tag{4.24}$$

and we now claim that $h_1 | d_1$. For if not, we could find integers q and r such that $d_1 = q h_1 + r$, where $0 < r < h_1$. Thus $y - q y_1 = r v_1 + d_2 v_2' + \ldots + d_n v_n'$ would be an element of H such that $\delta(y - q y_1) = (r, d_2, \ldots, d_n) \leqq r < h_1$, contradicting the minimality of h_1. Hence we conclude that $r = 0$, that is

$$y - q y_1 = d_2 v_2' + \ldots + d_n v_n'. \tag{4.25}$$

(iii) The proof proceeds by induction on n. When $n = 1$, we have already reached our goal. For in that case, the right-hand side of

(4.25) must be replaced by zero, and $y = qy_1 = qh_1v_1$. This amounts to the statement that $F = \langle v_1 \rangle$, $H = \langle h_1v_1 \rangle$, as asserted by the theorem when $n = m = 1$. Now suppose that $n > 1$ and put

$$F_1 = \langle v_2', v_3', \ldots, v_n' \rangle, \quad H_1 = H \cap F_1. \qquad (4.26)$$

Note that the right-hand side of (4.25) belongs to F_1 whilst the left-hand side lies in H. Hence (4.25) represents an element of H_1. Two cases have to be considered: first, if $H_1 = \{0\}$, we have that $y = qy_1 = qh_1v_1$ and, as before $H = \langle h_1v_1 \rangle$. This, together with (4.23), establishes the theorem, when n is arbitrary and $m = 1$. Next, when H_1 is a non-zero subgroup of F_1 we apply the inductive hypothesis to F_1 and H_1. Thus we can find elements $v_2, v_3, \ldots, v_n$ of F_1 such that

$$F_1 = \langle v_2, v_3, \ldots, v_n \rangle, \quad H_1 = \langle h_2v_2, h_3v_3, \ldots, h_mv_m \rangle, \quad (4.27)$$

where m is a certain integer satisfying $2 \leq m \leq n$ and $h_i|h_{i+1}$ $(i = 2, 3, \ldots, m-1)$. The two sets of free generators for F_1 are linked by equations

$$v_i = \sum_j p_{ij}v_j' \quad v_i' = \sum_j q_{ij}v_j \quad (i, j = 2, 3, \ldots, n).$$

We claim that

$$F = \langle v_1, v_2, \ldots, v_n \rangle. \qquad (4.28)$$

For on expressing the v's in terms of v''s in (4.23) we see that $v_1, v_2, \ldots, v_n$ certainly generate F. Moreover, these elements are free generators; for suppose that we have a non-trivial relation

$$c_1v_1 + c_2v_2 + \ldots + c_nv_n = 0. \qquad (4.29)$$

We observe that $c_1 \neq 0$; for if not, we should have a relation between $v_2, v_3, \ldots, v_n$ in contradiction to (4.27). If we now substitute in (4.29) for $v_2, v_3, \ldots, v_n$ in terms of $v_2', v_3', \ldots, v_n'$, we should get a relation between $v_1, v_2', \ldots, v_n'$ in which the coefficient of v_1 is c_1. This is incompatible with (4.23). Thus (4.28) is established. Combining (4.24), (4.25) and (4.27), we find that the elements

$$h_1v_1(= y_1), h_2v_2, \ldots, h_mv_m$$

generate H. In fact, they are free generators, because any non-trivial relation between them would also be a relation between $v_1, v_2, \ldots, v_m, \ldots v_n$ and would therefore contradict (4.28). Thus

$$H = \langle h_1v_1, h_2v_2, \ldots, h_mv_m \rangle.$$

D

To conclude the proof we have still to show that $h_1|h_2$. Now $y_0 = h_1v_1 + h_2v_2$ is an element of H. Hence by the minimality of h_1, $\delta(y_0) = (h_1, h_2) \geq h_1$. From the definition of a HCF, $(h_1, h_2) \leq h_1$. Therefore $(h_1, h_2) = h_1$, that is $h_1|h_2$.

27. Finitely generated Abelian Groups. We now turn to the discussion of an arbitrary f.g. Abelian group A. Of course, all finite Abelian groups belong to this class. Let

$$A = \text{gp} \{s_1, s_2, \ldots, s_n\},$$

where it is now admitted that the generators $s_1, s_2, \ldots, s_n$ may satisfy non-trivial relations. We associate with A the free Abelian group

$$F = \langle u_1, u_2, \ldots, u_n \rangle,$$

which is freely generated by the symbols $u_1, u_2, \ldots, u_n$. In order to establish a connection between A and F we introduce the map

$$\theta: F \to A,$$

defined by

$$(a_1u_1 + a_2u_2 + \ldots + a_nu_n)\theta = a_1s_1 + a_2s_2 + \ldots + a_ns_n.$$

$$(4.30)$$

We leave to the reader the easy verification that θ is indeed a homomorphism. Evidently, θ is surjective, because any element of A can appear on the right-hand side of (4.30). Let R be the kernel of θ; this is a subgroup of F. Thus the element $a_1u_1 + a_2u_2 + \ldots + a_nu_n$ of F belongs to R if and only if $a_1s_1 + a_2s_2 + \ldots + a_ns_n = 0$. This is a relation between the generators of A, and we can say that the elements of R are in one-to-one correspondence with all relations satisfied by the generators of A. The First Isomorphism (Theorem (p. 68)) now states that

$$A \cong F/R, \qquad (4.31)$$

and we are able to discover the structure of A by examining F/R, for which we are well equipped by the preceding section. Thus we may choose free generators $v_1, v_2, \ldots, v_n$ for F in such a way that

$$F = \langle v_1, v_2, \ldots, v_n \rangle, \quad R = \langle h_1v_1, h_2v_2, \ldots, h_mv_m \rangle \quad (4.32)$$

$(m \leq n)$, and $h_i|h_{i+1}(i = 1, 2, \ldots, m-1)$, provided that $R \neq \{0\}$.

By way of preparation, let us consider the case in which $n = 1$. Three cases have to be distinguished:

(i) $F = \langle v \rangle$, $R = \{0\}$. Then $F/R \cong F$, the infinite cyclic group generated by v.

(ii) $F = \langle v \rangle$, $R = \langle hv \rangle$, where $h \geq 2$. Then $F/R \cong C_h$ the cyclic group of order h.

(iii) $F = \langle v \rangle$, $R = \langle v \rangle$ ($h = 1$). Then $F/R \cong \{0\}$, because $F = R$.

In the general situation the same features will appear, and it is desirable to adapt the notation to the three types of generators. If $r = n - m > 0$, there are r generators in F which do not occur in R; these will be denoted by $x_1, x_2, \ldots, x_r$. If $h_1 = h_2 = \ldots = h_l = 1$, the corresponding generators, say $z_1, z_2, \ldots, z_l$ will occur both in F and in R. If $n = r + l + k$, the remaining k generators correspond to values of h which are greater than one, and it is convenient to rearrange them in decreasing order of magnitude, say $e_1, e_2, \ldots, e_k$. Thus we shall write

$$F = \langle x_1, x_2, \ldots, x_r, y_1, y_2, \ldots, y_k, z_1, z_2, \ldots, z_l \rangle, \quad (4.33)$$

$$R = \langle e_1 y_1, e_2 y_2, \ldots, e_k y_k, z_1, z_2, \ldots, z_l \rangle, \quad (4.34)$$

where $e_{\kappa+1} | e_\kappa$ ($\kappa = 1, 2, \ldots, k-1$), $n = r + k + l$, $m = k + l$, with the obvious modifications when one or the other type does not occur.

If $x \in F$, let $\bar{x} = x + R$ be the image of x under the natural epimorphism $F \to F/R$. In particular, considering the generators of F in turn, we find that (i) $\bar{x}_\rho (\rho = 1, 2, \ldots, r)$ is an element of infinite order because no multiple ($\neq 0$) of x_ρ lies in R; (ii) $\bar{y}_\kappa$ is of order $e_\kappa (\kappa = 1, 2, \ldots, k)$; (iii) $\bar{z}_\lambda (\lambda = 1, 2, \ldots, l)$ is the zero element $\bar{0}$ of F/R because $z_\lambda \in R$. Now the general element of F can be expressed in the form

$$x = \sum_{\rho=1}^{r} a_\rho x_\rho + \sum_{\kappa=1}^{k} b_\kappa y_\kappa + \sum_{\lambda=1}^{l} c_\lambda z_\lambda,$$

whence a typical element of F/R becomes

$$\bar{x} = \sum_{\rho=1}^{r} a_\rho \bar{x}_\rho + \sum_{\kappa=1}^{k} b_\kappa \bar{y}_\kappa. \quad (4.35)$$

Thus F/R is generated by $\bar{x}_1, \ldots, \bar{x}_r, \bar{y}_1, \ldots, \bar{y}_k$. We claim, however, that in fact

$$F/R = \mathrm{gp}\,\{\bar{x}_1\} \oplus \ldots \oplus \mathrm{gp}\,\{\bar{x}_r\} \oplus \mathrm{gp}\,\{\bar{y}_1\} \oplus \ldots \oplus \mathrm{gp}\,\{\bar{y}_k\},$$

$$(4.36)$$

that is we assert that the right-hand side of (4.35) can vanish only if each term is zero. Suppose that

$$\sum_{\rho=1}^{r} a_\rho \bar{x}_\rho + \sum_{\kappa=1}^{k} b_\kappa \bar{y}_\kappa = \bar{0}.$$

This means that

$$\sum_{\rho=1}^{r} a_\rho x_\rho + \sum_{\kappa=1}^{k} b_\kappa y_\kappa \in R.$$

A glance at (4.34) shows that $a_\rho = 0$ ($\rho = 1, 2, \ldots, r$) because x_ρ does not occur in R. Furthermore, b_κ must be divisible by e_κ ($\kappa = 1, 2, \ldots, k$), say $b_\kappa = d_\kappa e_\kappa$. Hence $b_\kappa \bar{y}_\kappa = d_\kappa e_\kappa \bar{y}_\kappa = \bar{0}$, because $e_\kappa \bar{y}_\kappa = \bar{0}$. This proves (4.36). Since, by (4.31) we may identify the given group A with F/R, we have established the following fundamental theorem:

THEOREM 13. (Basis Theorem for f.g. Abelian groups). *Every f.g. Abelian group A is the direct sum of cyclic groups, involving $r(\geqq 0)$ infinite and $k(\geqq 0)$ finite cyclic groups; thus*

$$A = \text{gp}\{t_1\} \oplus \ldots \oplus \text{gp}\{t_r\} \oplus \text{gp}\{w_1\} \oplus \ldots \oplus \text{gp}\{w_k\}, \quad (4.37)$$

where $t_\rho(\rho = 1, 2, \ldots, r)$ is of infinite order, whilst $w_\kappa(\kappa = 1, 2, \ldots, k)$ is of finite order $e_\kappa(\geqq 2)$. Moreover,

$$e_{\kappa+1}|e_\kappa \quad (\kappa = 1, 2, \ldots, k-1). \quad (4.38)$$

This theorem effectively solves the problem of describing the structures of all f.g. Abelian groups. The generators which occur in the direct decomposition (4.37) are said to form a **basis** of A. To repeat, they are not free or 'independent', like the basis elements of a vector space, but have the property that in a non-trivial relation each term is zero.

When $r = 0$, the group A is finite and $|A| = e_1 e_2 \ldots e_k$; in the other extreme case, when $k = 0$, A is a free Abelian group. The number, r, of free generators is called the **rank** of A, whether or not A is free.

The decomposition described in Theorem 13 is called a **canonical form** for A. This, somewhat vague, term is used whenever the structure of a mathematical object has been displayed in a simple and essentially unique manner. The question of uniqueness, which we have so far left aside, will be discussed in the next section.

28. Invariants and Elementary Divisors. The uniqueness just mentioned is made precise as follows:

THEOREM 14. *Let A be a f.g. Abelian group and suppose that*

$$A = \text{gp} \{x_1\} \oplus \ldots \oplus \text{gp} \{x_r\} \oplus \text{gp} \{u_1\} \oplus \ldots \oplus \text{gp} \{u_k\}$$
$$\text{(4.39)}$$

$$= \text{gp} \{y_1\} \oplus \ldots \oplus \text{gp} \{y_s\} \oplus \text{gp} \{v_1\} \oplus \ldots \oplus \text{gp} \{v_l\},$$
$$\text{(4.40)}$$

where $x_\rho(\rho = 1, 2, \ldots, r)$ *and* $y_\sigma(\sigma = 1, 2, \ldots, s)$ *are elements of infinite order, and* $|u_\kappa| = d_\kappa(\kappa = 1, 2, \ldots, k)$, $d_{\kappa+1}|d_\kappa$, $|v_\lambda| = e_\lambda$ $(\lambda = 1, 2, \ldots, l)$, $e_{\lambda+1}|e_\lambda$. *Then* (i) $r = s$ *and* (ii) $k = l$, $d_\kappa = e_\kappa$ $(\kappa = 1, 2, \ldots, k)$.

The proof of this theorem will occupy most of the section and will be divided into several stages.

(i) Let T be the collection of those elements of A which are of finite order. If $u, v \in T$, there exist positive integers m and n such that $mu = nv = 0$. Hence $mn(u-v) = 0$ so that $u-v \in T$. It follows that $u-v \in T$, and this proves that T is a subgroup (see p. 82). This group is called the **torsion subgroup** of A, a term borrowed from topology. Of course T is intrinsically related to A, that is, it does not depend on the choice of basis elements. Now

$$X = \sum_{\rho=1}^{r} \oplus \text{gp} \{x_\rho\} \quad \text{and} \quad Y = \sum_{\sigma=1}^{s} \oplus \text{gp} \{y_\sigma\}$$

are free Abelian groups of ranks r and s respectively. The hypotheses (4.39) and (4.40) imply that

$$A = X \oplus T = Y \oplus T. \tag{4.41}$$

For it is clear that the torsion group cannot involve any generator of infinite order whilst it necessarily contains all generators of finite order. We deduce from (4.41) that $A/T \cong X$, $A/T \cong Y$, whence $X \cong Y$. But the rank of a free Abelian group is an invariant (p. 85). Hence $r = s$, and the first part of Theorem 14 is established.

(ii) From now on we shall be concerned exclusively with finite Abelian groups, that is, we ignore the infinite generators in (4.39) and (4.40). We begin with the special case in which A is a finite Abelian p-group, that is we suppose that $|A| = p^m$, where p is a prime and m is a positive integer. Then the order of each element is a power of p, and in particular we put

$$|u_\kappa| = d_\kappa = p^{\delta_\kappa} \ (\kappa = 1, 2, \ldots, k), |v_\lambda| = e_\lambda = p^{\varepsilon_\lambda} \ (\lambda = 1, 2, \ldots, l).$$

The condition $d_{\kappa+1} | d_\kappa$ is equivalent to $\delta_{\kappa+1} \leq \delta_\kappa$; similarly $\varepsilon_{\lambda+1} \leq \varepsilon_\lambda$. The adaptation of Theorem 14 to p-groups therefore amounts to the following statement.

THEOREM 15. *Let A be a finite Abelian p-group. Assume that*

$$A = \sum_{\kappa=1}^{k} \oplus \text{ gp } \{u_\kappa\} = \sum_{\lambda=1}^{l} \oplus \text{ gp } \{v_\lambda\}, \qquad (4.42)$$

where $|u_\kappa| = p^{\delta_\kappa}$ ($\kappa = 1, 2, \ldots, k$), $|v_\lambda| = p^{\varepsilon_\lambda}$ ($\lambda = 1, 2, \ldots, l$) and $\delta_1 \geq \delta_2 \geq \ldots \geq \delta_k, \varepsilon_1 \geq \varepsilon_2 \geq \ldots \geq \varepsilon_l$. Then $k = l$ and $\delta_\kappa = \varepsilon_\kappa$ ($\kappa = 1, 2, \ldots, k$).

*Proof.** If $|A| = p^m$, then on comparing the orders in (4.42),

$$m = \sum_\kappa \delta_\kappa = \sum_\lambda \varepsilon_\lambda.$$

When $m = 1$, the theorem is trivial. We may therefore proceed by induction on m.

Let A_p be the set of elements satisfying $px = 0$. Since $p(x-y) = px - py$, it follows that A_p is a subgroup of A (possibly equal to A). It is easy to determine the order of A_p. For let $x \in A_p$. Using the basis $u_1, u_2, \ldots, u_k$ for A we have that

$$x = \sum_{i=1}^{k} a_i u_i,$$

where it may be assumed that $0 \leq a_i < p^{\delta_i}$, because $|u_i| = p^{\delta_i}$. Now if $px = 0$, then $pa_i u_i = 0$ for each i and hence $p^{\delta_i} | pa_i$. Thus $a_i = b_i p^{\delta_i - 1}$, where b_i must satisfy $0 \leq b_i < p$. Hence for a fixed i, there are precisely p possible values for b_i and hence for a_i such that $px = 0$. This shows that $|A_p| = p^k$. Using the second basis in (4.42) we find analogously that $|A_p| = p^l$. But A_p is independent of the basis. Therefore $k = l$, as claimed.

Next, we define the set A^p consisting of all those elements x of A which are the pth multiple of some other element (the additive analogue of being a pth power). It is easily verified that A^p is in fact a group; for if $x = px'$, $y = py'$, then $x - y = p(x' - y')$. We obtain a direct decomposition of A^p into cycles if we simply multiply each basis element of A by p. But it must be observed that this operation 'kills' all elements of order p. Thus let

$$\delta_1 \geq \delta_2 \geq \ldots \geq \delta_K > 1, \delta_{K+1} = \delta_{K+2} = \ldots = \delta_k = 1,$$

---

* We follow Marshall Hall Jr., *The Theory of Groups* (New York, 1959), p. 41.

where K is a certain integer satisfying $0 \leq K \leq k$. Then

$$A^p = \sum_{i=1}^{K} \oplus \text{ gp } \{pu_i\} \quad \text{and} \quad |pu_i| = p^{\delta_i - 1}.$$

Similarly, if $\varepsilon_1 \geq \varepsilon_2 \geq \ldots \geq \varepsilon_L > 1$, $\varepsilon_{L+1} = \varepsilon_{L+2} = \ldots = \varepsilon_k = 1$, we can write

$$A^p = \sum_{j=1}^{L} \oplus \text{ gp } \{pv_j\},$$

where $|pv_j| = p^{\varepsilon_j - 1}$. When $K = 0$, all elements of A are of order p, whence $A^p = \{0\}$. In that case also $L = 0$, because A^p is independent of the basis. Henceforth, we shall assume that $K > 0$. Evidently, $|A^p| < |A|$, and we may apply the inductive hypothesis to A^p. Thus we conclude that $K = L$ and $\delta_i - 1 = \varepsilon_i - 1$, that is $\delta_i = \varepsilon_i (i = 1, 2, \ldots, K)$. Since the remaining δ's and ε's are equal to one, the proof of the theorem is complete.

The invariants

$$p^{\delta_1}, p^{\delta_2}, \ldots, p^{\delta_k} \tag{4.43}$$

of an Abelian p-group A are also called the **elementary divisors** of A. Thus two Abelian p-groups are isomorphic if and only if they have the same elementary divisors, arranged in some order. When the value of p is understood, it suffices to name the exponents in (4.43), and we say that A is of **type** $(\delta_1, \delta_2, \ldots, \delta_k)$. In particular, A is called an elementary Abelian p-group if it is of type $(1, 1, \ldots, 1)$.

(iii) The next step consists in breaking up an arbitrary finite Abelian group into p-groups. We begin with a lemma which is concerned with a single cyclic subgroup and, in its multiplicative version, would apply also to non-Abelian groups (see example 8, Chapter I).

LEMMA. *Let w be an element of order mn, where $(m, n) = 1$. Then*

$$\text{gp } \{w\} = \text{gp } \{nw\} \oplus \text{gp } \{mw\} \tag{4.44}$$

Proof. The elements $u = nw$ and $v = mw$ are of orders m and n respectively. Put $W = \text{gp } \{w\}$, $U = \text{gp } \{u\}$, $V = \text{gp } \{v\}$. We claim that

$$W = U \oplus V. \tag{4.45}$$

Since $(m, n) = 1$, there exist integers a and b such that $an + bm = 1$. Hence

$$w = (an + bm)w = a(nw) + b(mw)$$
$$= au + bv.$$

This shows that $w \in U + V$. But w generates W, whence $W \subset U + V$. Since $U \subset W$ and $V \subset W$, we have, conversely, that $U + V \subset W$ and therefore $W = U + V$. In order to prove that the sum is direct, we observe that $U \cap V = \{0\}$, since U and V are groups of coprime orders.

The result (4.44) can be generalized to more than two terms. In particular, let

$$m = p_1^{\alpha_1} p_2^{\alpha_2} \ldots p_t^{\alpha_t},$$

where $p_1, p_2, \ldots, p_t$ are distinct primes. Then

$$\text{gp } \{w\} = \sum_{\tau=1}^{t} \oplus \text{ gp } \{w_\tau\}, \qquad (4.46)$$

where $w_\tau = (m/p_\tau^{\alpha_\tau})w$ is of order $p_\tau^{\alpha_\tau}$. The formula (4.46) can still be used when some of the α's are zero. The corresponding summand then reduces to the zero group and can be omitted.

Let p be a prime and let P be the set of those elements of A whose order is a power of p, that is which satisfy an equation of the form $p^\mu x = 0$ ($\mu \geq 0$). Evidently, P is a subgroup; for if $p^\mu x = p^\nu y = 0$, then $p^{\mu+\nu}(x-y) = 0$. If p does not divide $|A|$, then $P = \{0\}$. We call P the p-**primary component** of A. Next, we show that when $|A|$ is divisible by more than one prime, the primary components furnish a decomposition of A.

THEOREM 16. *Let* $|A| = p_1^{v_1} p_2^{v_2} \ldots p_n^{v_n}$, *and let* P_i *be the* p_i-*primary component of* A. *Then*

$$A = P_1 \oplus P_2 \oplus \ldots \oplus P_n. \qquad (4.47)$$

Proof. If w is any element of A, then (4.46) shows that $w \in P_1 + P_2 + \ldots + P_n$ and therefore $A \subset P_1 + P_2 + \ldots + P_n$. Conversely, each P_i is contained in A, whence $A = P_1 + P_2 + \ldots + P_n$. Moreover, the sum is direct because the terms have mutually coprime orders (see (3), p. 43). The decomposition (4.47) is unique in the following sense: let

$$A = P_1^* \oplus P_2^* \oplus \ldots \oplus P_n^*,$$

where P_i^* is an Abelian p_i-group ($i = 1, 2, \ldots, n$). Then $P_i^* = P_i$. For let $|P_i^*| = p_i^{\mu_i}$; computing the order of the group on each side of (4.47) we find that $|A| = \Pi_i p_i^{\mu_i}$, whence, by the unique factorization of $|A|$ into prime factors, it follows that $\mu_i = v_i$. Thus $|P_i^*| = |P_i|$. Now by the definition of P_i each element of P_i^* lies in

P_i, that is $P_i^* \subset P_i$. Since these two groups have the same order, we conclude that $P_i^* = P_i$.

(i) At last, we return to the proof of Theorem 14. The notation of Theorem 16 is retained. We are given that

$$A = \sum_{\kappa=1}^{k} \oplus \text{gp}\{u_\kappa\}, \quad |u_\kappa| = d_\kappa, \quad d_{\kappa+1}|d_\kappa. \tag{4.48}$$

The idea of the proof is to break up each term into its primary components and thence to obtain the elementary divisors of $P_1, P_2, \ldots, P_n$, whose uniqueness has been established in Theorem 15. Let

$$d_\kappa = \prod_{i=1}^{n} p_i^{\delta_{\kappa i}} \quad (\kappa = 1, 2, \ldots, k), \tag{4.49}$$

where $\delta_{\kappa i} \geq 0$ and $\delta_{\kappa+1,i} \leq \delta_{\kappa i}$. Applying (4.46) to each u_κ in turn we can write

$$\text{gp}\{u_\kappa\} = \sum_{i=1}^{n} \oplus \text{gp}\{u_{\kappa i}\},$$

where $|u_{\kappa i}| = p^{\delta_{\kappa i}}$. Thus we can express A as a double sum of p-groups, namely

$$A = \sum_{\kappa=1}^{k} \sum_{i=1}^{n} \oplus \text{gp}\{u_{\kappa i}\}. \tag{4.50}$$

For fixed i, we find that

$$P_i = \sum_{\kappa=1}^{k} \oplus \text{gp}\{u_{\kappa i}\}.$$

This shows that the elementary divisors of P_i are the non-units among the monotonic decreasing sequence

$$p^{\delta_{1i}}, p^{\delta_{2i}}, \ldots, p^{\delta_{ki}}.$$

The situation is summarized by the following table, in which prime powers are simply indicated by their exponents

	p_1	p_2	$\cdots$	p_n
d_1	δ_{11}	δ_{12}	$\cdots$	δ_{1n}
d_2	δ_{21}	δ_{22}	$\cdots$	δ_{2n}
.	.	.		.
.	.	.		.
.	.	.		.
d_k	δ_{k1}	δ_{k2}	$\cdots$	δ_{kn}

$$(4.51)$$

The rows correspond to (4.49), whilst the non-zero elements in the columns determine the elementary divisors of $P_1, P_2, \ldots, P_n$. The entries in each column are arranged in non-increasing magnitude, and the last row is not entirely zero, because $d_k \geq 2$.

Now suppose we replace the d's by a rival set of e's, where

$$e_\lambda = \prod_{i=1}^{n} p^{\varepsilon_{\lambda j}}. \quad (\lambda = 1, 2, \ldots, l).$$

Theorem 15 ensures that in the tables $(\delta_{\kappa i})$ and $(\varepsilon_{\lambda i})$ corresponding columns have the same non-zero entries. Since at least one column of $(\delta_{\kappa i})$ has k non-zero entries, it follows that $l \geq k$ similarly, by symmetry, $k \geq l$. Hence $k = l$ and the tables $(\delta_{\kappa i})$ and $(\varepsilon_{\lambda i})$ are identical. This concludes the proof of Theorem 14. The integers $d_1, d_2, \ldots, d_k$ are called the **invariants** of A; they are always assumed to satisfy the divisibility condition $d_{\kappa+1} | d_\kappa$. The **elementary divisors** of A are the collection of elementary divisors of the primary components $P_i (i = 1, 2, \ldots, n)$. The foregoing argument shows that the invariants and elementary divisors determine each other. Either set completely describes the structure of A, and all f.g. Abelian groups are obtained up to isomorphism by prescribing either the invariants or the elementary divisors. The latter leads to the decomposition (4.50) involving the greatest number of cyclic summands, whilst the former yield the decomposition (4.48) with the smallest number of terms.

Example 1. Find the invariants of the group whose elementary divisors are 2^3, 2, 2, 3, 3. The table (4.51) becomes

	2	3
d_1	3	1
d_2	1	1
d_3	1	0

whence $d_1 = 2^3 \times 3 = 24$, $d_2 = 2 \times 3 = 6$, $d_3 = 2$. The order of the group is

$$|A| = 24 \times 6 \times 2 = 2^5 \times 3^2 = 288.$$

The next example illustrates how a direct sum of cyclic groups can be turned into either of the two canonical forms which correspond to the elementary divisors or invariants respectively.

Example 2. Find the elementary divisors and invariants of the group

$$A = C_{30} \oplus C_{12}.$$

This is not in canonical form, because $12 \nmid 30$. First, we split each term into groups of coprime orders, thus

$$A = (C_2 \oplus C_3 \oplus C_5) \oplus (C_4 \oplus C_3).$$

On collecting terms belonging to the same prime we have that

$$A = (C_4 \oplus C_2) \oplus (C_3 \oplus C_3) \oplus C_5.$$

This shows that the elementary divisors which correspond to the primes 2, 3 and 5 are (4, 2), (3, 3) and 5 respectively. In fact, the scheme (4.51) becomes

	2	3	5
d_1	2	1	1
d_2	1	1	0

whence $d_1 = 2^2 \times 3 \times 5 = 60$, $d_2 = 2 \times 3 = 6$. Thus

$$A = C_{60} \oplus C_6$$

is the canonical form exhibiting the invariants.

29. Technique of Decomposition. In §27 we established the fundamental result that every f.g. Abelian group A is isomorphic with a direct sum of cyclic groups. But the arguments used in the proof do not immediately lead to a practical method for determining the cyclic summands. The aim of the present section is to describe a systematic procedure for solving the problem in concrete cases.

Suppose that A is given in terms of generators and relations. Thus we assume that

$$A = \text{gp}\{x_1, x_2, \ldots, x_n\},$$

where the generators $x_1, x_2, \ldots, x_n$ are subject to the N relations

$$\sum_{j=1}^{n} b_{ij}x_j = 0 \quad (i = 1, 2, \ldots, N).$$

The $N \times n$ integral matrix $B = (b_{ij})$ will be called the *relation matrix*. As in §27 we reformulate the problem by introducing a free Abelian group

$$F = \langle u_1, u_2, \ldots, u_n \rangle \tag{4.52}$$

and a relation subgroup

$$R = \text{gp } \{r_1, r_2, \ldots, r_N\}, \qquad (4.53)$$

where

$$r_i = \sum_{j=1}^{n} b_{ij} u_j \quad (i = 1, 2, \ldots, N).$$

It should be noted that the u_j are by definition free generators of F, whilst the r_i merely generate R. As we have seen in (4.31), the group A then appears in the form F/R, and its structure is revealed when new generators have been chosen in such a way that (4.32) is satisfied. Relative to these generators the relation matrix has the simple property that all non-diagonal elements are zero. Conversely, if

$$B = \begin{pmatrix} d_1 & 0 & 0 & \cdots \\ 0 & d_2 & 0 & \cdots \\ 0 & 0 & d_3 & \cdots \end{pmatrix}, \qquad (4.54)$$

the decomposition of Z into cyclic groups can be read off. However, this decomposition does not agree with the canonical form described in Theorem 13, unless the further conditions $d_{i+1} | d_i$ are fulfilled. For technical reasons it is advantageous to ignore these conditions in the first instance and to aim at a provisional reduction which corresponds to a relation matrix which is merely diagonal (see Example 2, p. 99).

The problem may be set out in tabular form as follows

	u_1	u_2	$\cdots$	u_n	
r_1	b_{11}	b_{12}	$\cdots$	b_{1n}	
r_2	b_{21}	b_{22}	$\cdots$	b_{2n}	
.	.	.		.	(4.55)
.	.	.		.	
.	.	.		.	
r_N	b_{N1}	b_{N2}		b_{Nn}	

The columns in this scheme correspond to the generators of F, whilst the rows exhibit the generators of R: since the choice of generators, both for F and for R, is at our disposal, we may apply the operations (α), (β), (γ), (δ) (p. 82) to each set of generators without changing the structure of F/R. As regards R, this amounts to operating on the rows of the scheme (4.55), but a little more care is needed to understand the effect of modifying the generators of F. Suppose we wish to introduce new generators for F by means of the transformation

$$u_1' = u_1 + qu_2, u_2' = u_2, u_3' = u_3, \ldots, u_n' = u_n, \qquad (4.56)$$

where q is an arbitrary integer, and let

$$r = b_1 u_1 + b_2 u_2 + \ldots + b_n u_n$$

be a typical element of the relation subgroup. When referred to the new generators this relation becomes

$$r = b_1 u_1' + (b_2 - qb_1)u_2' + b_3 u_3' + \ldots + b_n u_n'.$$

Thus the top row of (4.55) is replaced by (4.56) whilst, in the matrix B, q times the first column is subtracted from the second column. This is a (β)-type operation on the columns.

We shall now indicate a sequence of steps which will reduce B to the diagonal form (4.54).

(i) When $B = 0$, $A = F$ is a free Abelian group, and there is no more to say. Hence we shall now assume that $B \neq 0$. By a permutation of the rows and of the columns, and, if necessary, a change of sign in one of them, we can arrange that the 'pivot' b_{11} satisfies

$$b_{11} > 0, b_{11} \leq |b_{i1}|, b_{11} \leq |b_{1j}| \quad (i > 1, j > 1).$$

(ii) It may happen that all elements of B which are vertically or horizontally aligned with b_{11}, are divisible by b_{11}. In that case we can reduce all these elements to zero by subtracting suitable multiples of the first row (column) from the other rows (columns). When this has been done, the relation matrix becomes

$$\begin{pmatrix} b_{11} & 0 \\ 0 & B_1 \end{pmatrix}, \qquad (4.57)$$

and we proceed by treating B_1 in the same way until (4.54) is reached.

(iii) If, on the other hand, one of the b_{i1} or b_{1j} is not divisible by b_{11}, a (β)-type operation will cause it to be replaced by its least positive remainder modulo b_{11}. For example, we may have that

$$b_{i1} - qb_{11} = b_{i1}',$$

where $0 < b_{i1}' < b_{11}$. We then bring b_{i1}' into the leading position and repeat the reduction with the new pivot. It is clear that this process must come to an end after a finite number of steps, since the pivotal position is occupied by a decreasing sequence of positive integers, so that eventually the situation described in (ii) must arise.

Example 3. Find the invariants of the Abelian group A which is generated by a, b and c subject to the relations

$$3a - 2b + 5c = 0, \quad 5a + 27c = 0.$$

The following sequence of operations on the relation matrix leads to the canonical form:

$$
\begin{matrix}
3 & -2 & 5 \\
5 & 0 & 27
\end{matrix}
\xrightarrow{(1)}
\begin{matrix}
1 & -2 & 5 \\
5 & 0 & 27
\end{matrix}
\xrightarrow{(2)}
\begin{matrix}
1 & -2 & 5 \\
0 & 10 & 2
\end{matrix}
$$

$$
\xrightarrow{(3)}
\begin{matrix}
1 & 0 & 0 \\
0 & 10 & 2
\end{matrix}
\xrightarrow{(4)}
\begin{matrix}
1 & 0 & 0 \\
0 & 0 & 2
\end{matrix}
\xrightarrow{(5)}
\begin{matrix}
1 & 0 & 0 \\
0 & 2 & 0
\end{matrix}
$$

We indicate the transitions to new generators or relations, which correspond to the various steps.

(1) generators $u_1 = u_1'$, $u_2 = u_1' + u_2'$, $u_3 = u_3'$.

(2) relations $r_1' = r_1$, $r_2' = r_2 - 5r_1$.

(3) generators $u_1' = u_1'' + 2u_2'' - 5u_3''$, $u_2' = u_2''$, $u_3' = u_3''$.

(4) generators $u_1'' = u_1'''$, $u_2'' = u_2'''$, $u_3'' = u_3''' - 5_2'''$.

(5) generators $u_1''' = v_1$, $u_2''' = v_3$, $u_3''' = v_2$.

This completes the reduction. Using the notation of p. 91 we find that F/R is generated by $\bar{v}_1, \bar{v}_2, \bar{v}_3$, where $\bar{v}_1 = 0$, $2\bar{v}_2 = 0$, whilst $\bar{v}_3$ is of infinite order.

Hence

$$A \cong C_2 \oplus C_\infty.$$

Eliminating the intermediate generators we have that

$$v_1 = 3u_1 - 2u_2 + 5u_3, \quad v_2 = 5u_1 + 5u_2 + u_3, \quad v_3 = -u_1 + u_2,$$

and the reader may verify that this is a unimodular transformation.

When it is considered inessential to record the change of generators, the cycle structure can be revealed by applying pivotal operations to the relation matrix until the diagonal form is obtained. In the following example, where column operations are sufficient, the jth column is denoted by c_j.

Example 3. Find the canonical decomposition of the Abelian group with generators a, b, c, d and relations

$$3a+9b-3c = 0, \ 4a+2b-2d = 0.$$

The relation matrix may be reduced as follows:

$$\begin{array}{cccc} 3 & 9 & -3 & 0 \\ 4 & 2 & 0 & -2 \end{array} \qquad \rightarrow \begin{array}{cccc} 3 & 9 & -3 & 0 \\ 0 & 0 & 0 & -2 \end{array}$$

$$(c_1 \rightarrow c_1+2c_4, \ c_2 \rightarrow c_2+c_4)$$

$$\rightarrow \begin{array}{cccc} 3 & 0 & 0 & 0 \\ 0 & 0 & 0 & 2 \end{array} \qquad \rightarrow \begin{array}{cccc} 3 & 0 & 0 & 0 \\ 0 & 2 & 0 & 0 \end{array}$$

$$(c_2 \rightarrow c_2-3c_1, \ c_3 \rightarrow c_3+c_1, \ c_4 \rightarrow -c_4) \qquad (c_2 \rightarrow c_4, \ c_4 \rightarrow c_2)$$

It follows that two generators remain free, whilst the other two correspond to cyclic groups of orders 3 and 2 respectively. Thus the group is isomorphic with

$$C_3 \oplus C_2 \oplus C_\infty \oplus C_\infty.$$

Exercises

(1) Prove that if $b_1, b_2, \ldots, b_n$ are integers such that $(b_1, b_2, \ldots, b_n) = 1$, there exists a unimodular matrix whose first row is $b_1, b_2, \ldots, b_n$.

(2) Show that if the order of a finite Abelian group is not divisible by a square (> 1), then the group is cyclic.

(3) Prove that in a finite Abelian group (i) the maximal order of an element is equal to the greatest invariant and (ii) the order of any element divides the maximal order.

(4) Show that the (multiplicative) group of residue classes coprime with 24 is elementary Abelian of order 8.

(5) Find the elementary divisors and invariants of the following Abelian groups defined by generators and relations: (i) $15a = 4b = 0$, (ii) $20a = 6b = 5c = 0$.

(6) The Abelian group A is generated by a, b, c with the defining relations $3a+9b+9c = 0$, $6a-12b = 0$. Express A as a direct sum of cyclic groups.

(7) Find the rank and invariants of the following Abelian groups: (i) with generators a, b and relation $2(a+b) = 0$; (ii) with generators a, b, c, d and relations $3a+5b-3c = 0$, $4a+2b-2d = 0$.

(8) The free Abelian group F is generated by u_1, u_2, u_3 and R is the subgroup generated by

$$r_1 = ku_1 + u_2 + u_3, \; r_2 = u_1 + ku_2 + u_3, \; r_3 = u_1 + u_2 + ku_3,$$

where k is an integer greater than one. Find generators v_1, v_2, v_3 of F and s_1, s_2, s_3 of R such that $s_i = e_i v_i (i = 1, 2, 3)$ and e_1, e_2, e_3 are integers satisfying $e_1 | e_2 | e_3$.

(9) Show that an Abelian group of order g has at least one subgroup whose order is equal to any pre-assigned factor of g. (*Converse of Lagrange's Theorem for Abelian groups.*)

(10) Verify that an elementary Abelian group of order p^k (p a prime) may be regarded as a k-dimensional vector space over the prime field consisting of the elements $0, 1, \ldots, p-1$.

(11) Show that in an elementary Abelian group of order p^3 an 'ordered' basis may be chosen in $p^3(p^3-1)(p^2-1)(p-1)$ ways. [Bases consisting of the same elements, but in a different order, are regarded as distinct.]

(12) Prove that the results of §27 yield the following fundamental theorem on matrices: *If B is an $m \times n$ integral matrix of rank k, there exist unimodular matrices P and Q of orders m and n respectively such that $PBQ = D$, where in $D = (d_{ij})$ all elements are zero except the first k diagonal elements and $d_{11} | d_{22} \ldots | d_{kk}$.*

V. Generators and Relations

30. Finitely generated and related Groups. In the preceding chapter
we saw that the structure of an Abelian group can be satisfactorily
determined provided that the group is generated by a finite number
of elements which are subject to a finite number of relations. The
question naturally arises whether an analogous theory applies to
non-Abelian groups. The problem was briefly mentioned in §12, and
we have met a few examples of non-Abelian groups which were
described in terms of generators and relations. As might be expected,
the absence of the commutative law renders the situation much more
complicated, and the scope of this book only allows us to present
the simplest ideas and facts of this extensive topic.

From the outset we shall confine ourselves to groups which by
hypothesis can be generated by a finite number of elements with a
finite number of relations. We shall say that such a group is *finitely
generated and related*.

31. Free Groups. We introduce non-commutative symbols $x_1, x_2,$
$\ldots, x_n$ with which we form **words**, that is formal products

$$w = x_a^{\alpha} x_b^{\beta} \ldots x_r^{\rho} \tag{5.1}$$

consisting of a finite number of factors. The suffixes $a, b, \ldots, r$ are
taken from the set of integers $1, 2, \ldots, n$, repetitions being allowed,
since the factors do not commute. The exponents $\alpha, \beta, \ldots, \rho$ are
positive or negative integers. We may regard a word as a function
of $x_1, x_2, \ldots, x_n$ and accordingly write w more explicitly as
$w(x_1, x_2, \ldots, x_n)$.

It is convenient to introduce the **empty word**, that is, a word in
which the number of factors is zero. The empty word will be denoted
by e, and we define

$$x_i^{0} = e \quad (i = 1, 2, \ldots, n).$$

A word is said to be **reduced**, if it is either the empty word or else if it is a product of the form (5.1) in which no two consecutive x's have the same suffix.

Multiplication of two non-empty words u and v is defined as follows: write down the formal product p consisting of the factors of u followed by those of v. If p happens to be a reduced word, we define it to be uv. In the contrary case, suppose that

$$u = u_0 x^\alpha, \qquad v = x^\beta v_0,$$

where x does not appear at the end of u_0 or at the beginning of v_0. We then simplify p by applying the rule

$$x^\alpha x^\beta = x^{\alpha+\beta}. \tag{5.2}$$

If $\alpha+\beta = 0$, the factor $x^{\alpha+\beta}$ is removed and further simplifications and cancellations may become possible. The process is continued until a reduced word p_0 is reached. We then define

$$uv = p_0.$$

It should be noted that the process of reduction is unique so that uv has an unambiguous meaning. The law of composition is supplemented by the obvious rule that

$$ue = eu = u,$$

that is the empty word acts as a unit element. The inverse of (5.1) is given by

$$w^{-1} = x_r^{-\rho} \ldots x_b^{-\beta} x_a^{-\alpha},$$

which is clearly a reduced word. The direct verification of the associative law

$$(uv)w = u(vw) \tag{5.3}$$

is somewhat laborious and is best carried out in several stages*, as follows:

(i) Let x be a single generator, and let u_0 and w_0 be reduced words (possibly empty) such that neither the last factor of u_0 nor the first factor of w_0 is a power of x with non-zero exponent. It is then readily seen that

$$(u_0 x^\alpha)(x^\beta w_0) = u_0(x^{\alpha+\beta} w_0) = (u_0 x^{\alpha+\beta})w_0.$$

(ii) If u and w are reduced words and x is any generator, then

$$(ux^\alpha)w = u(x^\alpha w). \tag{5.4}$$

* See A. G. Kurosh, 1955, *The theory of groups*, **1**, p. 126.

For let

$$u = u_0 x^\pi, \qquad w = x^\phi w_0,$$

where u_0 and w_0 are as in (i) and π and ϕ are integers, which may be zero. We then have that

$$
\begin{aligned}
(ux^\alpha)w &= [(u_0 x^\pi)x^\alpha](x^\phi w_0) = (u_0 x^{\pi + \alpha})(x^\phi w_0) \\
&= u_0(x^{\pi + \alpha + \phi} w_0) = u_0[x^\pi(x^{\alpha + \phi} w_0)] \\
&= u_0[x^\pi(x^\alpha w)] = (u_0 x^\pi)(x^\alpha w) = u(x^\alpha w).
\end{aligned}
$$

(iii) Finally, in order to prove (5.3) in general, we argue by induction on the number of factors in v. The case in which v reduces to a single factor x^α is covered by (5.4). Assume now that

$$v = v_0 x^\alpha$$

and that the associative law holds with v_0 in place of v. We then have that

$$
\begin{aligned}
(uv)w &= (uv_0 x^\alpha)w = [(uv_0)x^\alpha]w = (uv_0)(x^\alpha w) \\
&= u[v_0(x^\alpha w)] = u[(v_0 x^\alpha)w] = u(vw).
\end{aligned}
$$

This completes the verification of (5.3) in all cases.

The set of reduced words in the symbols $x_1, x_2, \ldots, x_n$ with the law of composition just defined is called the **free group** on $x_1, x_2, \ldots, x_n$. The free group on a single generator, x, is the infinite cyclic group (see §5). In the case of two generators, x and y, typical products are

$$(xy^{-2}x)(yx) = xy^{-2}xyz,$$

$$(xy^2)(y^{-1}x) = xyz,$$

$$(xyx^{-1})(xy^{-1}x) = x^2.$$

To summarize we may say that the free group on $x_1, x_2, \ldots, x_n$ consists of all reduced words in these symbols, and that these are subject only to the trivial conditions

$$x_i x_i^{-1} = x_i^{-1} x_i = e \quad (i = 1, 2, \ldots, n) \tag{5.5}$$

and their consequences. It should be noted that a free Abelian group on more than one generator is not a free group, because the nontrivial relation $xyx^{-1}y^{-1} = e$ holds in an Abelian group but not in a free group.

32. Relations. Suppose that G is a group which is generated by n of its elements, say

$$G = \text{gp}\{g_1, g_2, \ldots, g_n\}.$$

Then every element of G is a product of the form $g_a{}^\alpha g_b{}^\beta \ldots g_r{}^\rho$. Unless G is a free group, there are non-trivial equalities, such as

$$g_a{}^\alpha g_b{}^\beta \ldots = g_c{}^\gamma g_d{}^\delta \ldots,$$

or in more concise notation

$$r(g_1, g_2, \ldots, g_n) = 1, \tag{5.6}$$

where the left-hand side represents

$$(g_a{}^\alpha g_b{}^\beta \ldots)(g_c{}^\gamma g_d{}^\delta \ldots)^{-1}.$$

In order to analyse the situation in more detail we consider the free group F on n symbols $x_1, x_2, \ldots, x_n$ and then define the map

$$\theta : F \to G$$

of F onto G by the rule that

$$w(x_1, x_2, \ldots, x_n)\theta = w(g_1, g_2, \ldots, g_n), \tag{5.7}$$

that is, the image under θ of any product of the x's is the corresponding product of the g's; in particular,

$$e\theta = 1.$$

The important fact to note is that θ is a homomorphism. Thus if w_1 and w_2 are any elements of F, then

$$(w_1 w_2)\theta = (w_1\theta)(w_2\theta); \tag{5.8}$$

for $w_1 w_2$ is defined as the reduced word obtained by juxtaposition of w_1 and w_2 and subsequent simplification by virtue of the rules (5.2) and (5.5). But these rules hold in any group, and any operation carried out on the x_i is also valid for the g_i, and this is all that (5.8) means. In view of the fact that θ is a homomorphism we may specify it more simply by

$$x_i\theta = g_i \quad (i = 1, 2, \ldots, n), \tag{5.9}$$

whence (5.7) follows by repeated application. Let R be the kernel of θ, that is R consists of all those words $r(x_1, x_2, \ldots, x_n)$ of F which are mapped by θ into the left-hand sides of relations (5.6) in G. We recall that, by the First Isomorphism Theorem,

$$G \cong F/R. \tag{5.10}$$

Summarizing our results we can state the following theorem.

THEOREM 17. *Let F be the free group on $x_1, x_2, \ldots, x_n$. Then any group G which can be generated by n of its elements $g_1, g_2, \ldots, g_n$, is a homomorphic image of F by virtue of the map $x_i\theta = g_i (i = 1, 2, \ldots, n)$. The kernel of θ consists of all those words of F which become relations in G under the action of θ.*

The pair F, R of groups that appear on the right-hand side of (5.10) is said to form a **presentation** of G. A group may have many such presentations. Conversely, choose a normal subgroup R of F and define G as F/R. Then G has generators $g_i = x_iR$ ($i = 1, 2, \ldots, n$) and relations $r(g_1, g_2, \ldots, g_n) = 1$, where $r(x_1, x_2, \ldots, x_n)$ ranges over R; for $q(g_1, g_2, \ldots, g_n) = 1$ is a relation for G if and only if $q(x_1, x_2, \ldots, x_n)R = R$, that is

$$q(x_1, x_2, \ldots, x_n) \in R.$$

Thus the elements of R are in one-to-one correspondence with the relations satisfied by the generators of G. For this reason we shall call R the **relation group** of G.

33. Definition of a Group. We shall now discuss in more detail what is meant by saying that a group G is defined by n generators $g_1, g_2, \ldots, g_n$ and m relations

$$\rho_k(g_1, g_2, \ldots, g_n) = 1 \ (k = 1, 2, \ldots, m). \tag{5.11}$$

If $\sigma(g_1, g_2, \ldots, g_n) = 1$ and $\tau(g_1, g_2, \ldots, g_n) = 1$ are relations in G, then so are

$$\sigma(g_1, g_2, \ldots, g_n)\,\tau(g_1, g_2, \ldots, g_n) = 1,$$

$$\{\sigma(g_1, g_2, \ldots, g_n)\}^{-1} = 1$$

and

$$g^{-1}\{\sigma(g_1, g_2, \ldots, g_n)\}g = 1,$$

where g is an arbitrary element of G. Any relation, say

$$\rho(g_1, g_2, \ldots, g_n) = 1 \tag{5.12}$$

which is derived from the given relations (5.11) by applying the above operations any finite number of times, is called a **consequence** of (5.11).

Using the free group F on $x_1, x_2, \ldots, x_n$ we associate with the relation (5.12) the word $r = \rho(x_1, x_2, \ldots, x_n)$. Without loss of generality we may assume that this is a reduced word and therefore a legitimate element of F; for example we shall disallow the relation

$$g_1 g_2 g_2^{-2} g_1 g_2^{-1} = 1,$$

but replace it by

$$g_1 g_2^{-1} g_1 g_2^{-1} = 1.$$

The relations (5.11) correspond to the words

$$r_k = \rho_k(x_1, x_2, \ldots, x_n) \quad (k = 1, 2, \ldots, m). \tag{5.13}$$

These relations and all their consequences constitute the least normal subgroup R_0 of F which contains $r_1, r_2, \ldots, r_m$. This group is denoted by

$$R_0 = \{r_1, r_2, \ldots, r_m\}^F$$

and is called the **normal closure** of $r_1, r_2, \ldots, r_m$; it is precisely the subgroup of F generated by all the elements $w^{-1} r_k w$, where r_k ranges over $r_1, r_2, \ldots, r_m$ and w is an arbitrary element of F. Note that R_0 is determined completely by the set (5.11) and by F. By virtue of Theorem 17, G is isomorphic with F/R, where R is the relation group of G. Since R is the set of all words $r(x_1, x_2, \ldots, x_n)$ such that $r(g_1, g_2, \ldots, g_n) = 1$ is a relation for G, it follows that each element of R_0 belongs to R, that is

$$R_0 \leqq R. \tag{5.14}$$

We shall say that G is defined by the generators $g_1, g_2, \ldots, g_n$ and the relations (5.11), or more briefly that (5.11) is a *set of* **defining relations** *for G, if*

$$R_0 = R. \tag{5.15}$$

Expressed more informally, the condition (5.15) states that the given conditions (5.11) and their consequences contain every conceivable piece of information about the structure of G, provided that it is assumed from the outset that G is generated by n elements. Incidentally, we do not stipulate that either the generators or the list of relations should be irredundant. In most practical cases a small number of relations suffices to define a group. Nevertheless, unless

(5.11) is empty, the normal closure R_0 is an infinite group, which unfortunately is hard to compute as a rule, and indirect methods for describing G may have to be resorted to.

Next we have to consider the following existence problem: *given a set of relations* (5.11), *does there exist a group G on n generators for which* (5.11) *is a set of defining relations?* A simple construction demonstrates that the answer to this question is in the affirmative. Starting from (5.11) we form the normal closure R_0 and put

$$G_0 = F/R_0. \qquad (5.16)$$

This group is generated by the n cosets

$$g_i{}^0 = x_i R_0 \ (i = 1, 2, \ldots, n),$$

which satisfy all the relations (5.11). Indeed

$$\rho_k(g_i{}^0, g_2{}^0, \ldots, g_n{}^0) = \rho_k(x_1, x_2, \ldots, x_n)R_0 = r_k R_0 = R_0,$$

because $r_k \in R_0$. Suppose now that R is the relation group of G_0, defined at the end of §32 (p. 109). Then

$$G_0 \cong F/R.$$

If $r(x_1, x_2, \ldots, x_n)$ is an arbitrary element of R, it becomes a relation for G_0 when x_i is replaced by $g_i{}^0$ $(i = 1, 2, \ldots, n)$. Thus we have that

$$r(g_1{}^0, g_2{}^0, \ldots, g_n{}^0) = r(x_1, x_2, \ldots, x_n)R_0 = R_0,$$

whence $r \in R_0$. This means that $R \leq R_0$, which together with (5.14) implies (5.15). Hence (5.11) is a set of defining relations for G_0. When $R_0 = F$, only the trivial group satisfies (5.11).

The group G_0, which we have constructed, is the 'largest' or 'freest' group satisfying (5.11). This is made more precise by the following theorem.

THEOREM 18. *Let $G = \mathrm{gp}\{g_1, g_2, \ldots, g_n\}$ be a group with defining relations*

$$\rho_k(g_1, g_2, \ldots, g_n) = 1_G \ (k = 1, 2, \ldots, m). \qquad (5.17)$$

Suppose that $H = \mathrm{gp}\{h_1, h_2, \ldots, h_n\}$ satisfies the same relations, namely

$$\rho_k(h_1, h_2, \ldots, h_n) = 1_H \ (k = 1, 2, \ldots, m)$$

and possibly others which are not consequences of these. Then H is a homomorphic image of G by virtue of the map $\varepsilon : G \to H$, *given by*

$$g_i \varepsilon = h_i \quad (i = 1, 2, \ldots, n).$$

Proof. Since both G and H have n generators, there exist epimorphisms

$$\theta : F \to G, \qquad \eta : F \to H,$$

with kernels R and S, which are the relation groups of G and H respectively. Using the notation (5.14) we have that

$$R = R_0,$$

because (5.17) is a set of defining relations for G. The hypothesis about H is equivalent to the statement that

$$S \geqq R_0 (= R). \tag{5.18}$$

We now turn to the construction of the map

$$\varepsilon : G \to H.$$

(See Figure 2). Let $u = w(g_1, g_2, \ldots, g_n)$ be an arbitrary element

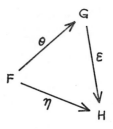

Fig. 2

of G. Since θ is an epimorphism, there exists an element z of F, for example $z = w(x_1, x_2, \ldots, x_n)$ (see (5.7)), such that

$$z\theta = u. \tag{5.19}$$

When u is given, the most general solution of (5.19) is zr, where r is any element of R. For $z\theta = z'\theta$, if and only if $z^{-1}z'$ belongs to R. We now claim that, if z satisfies (5.19), the equation

$$u\varepsilon = z\eta \tag{5.20}$$

determines a well-defined map ε from G onto H; that is, we must verify that the right-hand side of (5.20) remains unchanged if we replace z by zr. But

$$(zr)\eta = (z\eta)(r\eta) = z\eta,$$

because $r \in S$ by (5.18) and hence $r\eta = 1_H$. It is easy to check that

$$(u_1 \varepsilon)(u_2 \varepsilon) = (u_1 u_2)\varepsilon,$$

so that ε is indeed a homomorphism. In particular, when $u = g_i$, we may put $z = x_i$, and we find that

$$g_i \varepsilon = x_i \eta = h_i \quad (i = 1, 2, \ldots, n),$$

which clearly determines ε completely. This shows that ε is an epimorphism.

It is instructive to follow through the argument in the case of a group which we had previously (p. 40) studied informally. Let $G = \text{gp}\ \{a, c\}$ be then defined by the relations

$$a^3 = c^2 = (ac)^2 = 1, \tag{5.21}$$

as in (2.27). We use the free group F on the generators x and y and associate the elements

$$r_1 = x^3, \quad r_2 = y^2, \quad r_3 = (xy)^2$$

with the three relations given in (5.21). Let

$$R_0 = \{r_1, r_2, r_3\}^F$$

and $G_0 = F/R_0$. The elements of G_0 are the cosets wR_0, where $w \in F$, and it is easy to see that each coset is equal to one of

$$R_0, xR_0, x^2 R_0, yR_0, yxR_0, yx^2 R_0. \tag{5.22}$$

For example, $xyR_0 = yx^2 R_0$, because $xy = yx^2 r$, where

$$r = r_1^{-1}(xr_2^{-1}x)r_3$$

is an element of R_0. At this stage we cannot assert that the six cosets (5.22) are distinct; for there might be hidden consequences of (5.21) that render some of these cosets equal. However, $|G_0| \leq 6$, and we know that if H is any group on two generators which satisfy (5.21), then H is a homomorphic image of G and hence $|H| \leq |G_0|$. Now it happens that $H = S_3$ meets these requirements. For S_3 is generated by

$$\alpha = (1\ 2\ 3) \quad \text{and} \quad \gamma = (1\ 2),$$

and

$$\alpha^3 = \gamma^2 = (\alpha\gamma)^2 = \iota.$$

Since $|S_3| = 6$, we deduce that $|G_0| = 6$ and therefore $G_0 \cong S_3$.

As a further application of these ideas we mention the process of *making a group G Abelian*, that is, of passing from G to G/G', which is the largest Abelian homomorph of G. This amounts to adding the relations

$$g_i^{-1}g_j^{-1}g_ig_j = 1 \ (i < j)$$

to the existing relations. The structure of G/G' may then be found directly by the methods of Chapter IV.

Example: Find the structure of G/G' when G is the quaternion group

$$a^4 = 1, \quad a^2 = b^2, \quad ba = a^3b. \tag{5.17}$$

The Abelian group G/G' is generated by $\bar{a} = aG'$ and $\bar{b} = bG'$, which in the additive notation satisfy the following relations derived from (5.17):

$$4\bar{a} = 0, \quad 2\bar{a} = 2\bar{b}, \quad \bar{b}+\bar{a} = 3\bar{a}+\bar{b}.$$

These equations reduce to

$$2\bar{a} = 2\bar{b} = 0.$$

The corresponding relation matrix is already in diagonal form, whence we infer that

$$G/G' \cong C_2 \oplus C_2.$$

If the free group F on $x_1, x_2, \ldots, x_n$ is made Abelian in this way, we find that $F/F' = \langle \bar{x}_1, \bar{x}_2, \ldots, \bar{x}_n \rangle$ (see p. 84), where $\bar{x}_i = x_iF'$. Thus F/F' is a free Abelian group on n generators. Incidentally, it follows from this remark that free groups on different numbers of generators cannot be isomorphic. For let F_m and F_n be free groups on m and n generators respectively and suppose they are isomorphic. Then F_m/F_m' and F_n/F_n' would also be isomorphic. But these are free Abelian groups on m and n generators respectively, and we know that they cannot be isomorphic unless $m = n$ (p. 85). Finally, we mention without proof the important, but difficult theorem* which states that *every subgroup of a free group is a free group.*

---

* Marshall Hall, Jr., *loc. cit.*, p. 96.

Exercises

(1) Show that the derived group of a free group consists of those words in which the sum of the exponents for each generator is equal to zero (e.g. $x_1 x_2^{-1} x_1^{-2} x_2 x_1$).

(2) Determine the structure of G/G' when G is given by (i) $a^6 = b^2 = (ab)^2 = 1$; (ii) $a^6 = 1$, $b^2 = (ab)^2 = a^3$.

(3) Prove that if G is generated by a and b subject to the relations $a^{-1}ba = b^2$, $b^{-1}ab = a^2$, then $G = \{1\}$.

VI. Series of Subgroups

34. Nested Subgroups. It is common practice in mathematics to study complex entities by resolving them into simpler components which are 'irreducible'. Thus integers are factorized into primes, polynomials are split into irreducible factors, and so on. In order to be significant such a resolution must display features of uniqueness which correspond to intrinsic properties of the structure under investigation.

In the case of a group G the method consists in examining certain decreasing or increasing sequences of subgroups

$$A_1 \geqq A_2 \geqq \ldots \quad \text{or} \quad B_1 \leqq B_2 \leqq \ldots \qquad (6.1)$$

with appropriate additional properties.

Each of these nests of subgroups is designed to throw some light on the structure of G; but, as it turns out, none of them completely characterizes G. In this context the sequences (6.1) are referred to as *series* of subgroups (with apologies to Analysts).

35. The Jordan-Hölder Theorem. We recall that a group is called simple (p. 61) if it is of order greater than unity and has no non-trivial normal subgroup. For arbitrary groups the following definition describes an important type of normal subgroup.

DEFINITION 5: *A normal subgroup A ($\neq G$) is called a* **maximal normal subgroup** *of G, if there exists no normal subgroup H, other than G or A, such that*

$$G \rhd H \rhd A.$$

By Proposition 10 (p. 71), this is equivalent to the statement that G/A has no proper normal subgroup. Thus the above definition may be recast as follows.

CRITERION. *The normal subgroup A ($\neq G$) is a maximal normal subgroup of G, if and only if G/A is a simple group.*

It is possible for a group to possess several maximal normal subgroups differing both in structure and order. If G/A is of prime order, then A is a maximal normal subgroup. Again, if G is simple, then $\{1\}$ is the only maximal normal subgroup.

In order to simplify the discussion we shall confine ourselves to finite groups in the remainder of this section. The results hold also for certain classes of infinite groups (see e.g. [A. G. Kurosh, *loc. cit.*], **1**, 110–116); but the applications we have in mind are concerned with finite groups only. If G is not simple, let A_1 be one of its maximal normal subgroups; next, let A_2 be a maximal normal subgroup of A_1, let A_3 be a maximal normal subgroup of A_2, and so on. Since the groups we have defined are of strictly decreasing orders, we must eventually arrive at the unit group. Thus we are led to the following definition.

DEFINITION 6. *A sequence of subgroups*

$$A_1, A_2, \ldots, A_r \qquad (6.2)$$

of a group $G(=A_0)$ *is called a* **composition series** *of* G *if*

(i) $G \rhd A_1 \rhd A_2 \rhd \ldots \rhd A_r \rhd \{1\}$ \qquad (6.3)

and if

(ii) $G/A_1, A_1/A_2, \ldots, A_{r-1}/A_r, A_r$ \qquad (6.4)

are simple groups.

It must be clearly understood that whilst A_i is normal in A_{i-1}, and indeed maximal normal, it need not be normal in any other group preceding it in the sequence (6.3). In particular, among the groups (6.2) only A_1 is necessarily a normal subgroup of G. The quotient groups listed in (6.4) are called **composition quotient groups** or **composition factors**.

Since maximal normal subgroups are not, in general, unique, a group may possess more than one composition series. However, the following fundamental theorem asserts that the composition factors are unique, up to rearrangement and isomorphism. The set of composition factors therefore constitutes an intrinsic property of the group. We shall prove this result for finite groups only.

THEOREM 19. (JORDAN-HÖLDER). *In any two composition series of a finite group the composition factors are, apart from their sequence, isomorphic in pairs.*

Proof: Let us analyse this statement in more detail: suppose that

$$G\,(=A_0) \rhd A_1 \rhd A_2 \rhd \ldots \rhd A_r \rhd \{1\} \qquad (I)$$

and

$$G\,(=B_0) \rhd B_1 \rhd B_2 \rhd \ldots \rhd B_s \rhd \{1\} \qquad (II)$$

are two composition series for G. If it is the case that the composition factors

$$G/A_1, \quad A_1/A_2, \quad \ldots, A_{r-1}/A_r, A_r \qquad (I)'$$

and

$$G/B_1, \quad B_1/B_2, \quad \ldots, B_{s-1}/B_s, B_s \qquad (II)'$$

are isomorphic in pairs, apart from rearrangement, we shall write $(I) \sim (II)$. This clearly establishes an equivalence relation in the set of all possible composition series, and it is our aim to show that all composition series are equivalent in this sense. Notice that, in particular, $(I) \sim (II)$ implies that $r = s$.

When G is simple, the only possible composition series is $G \rhd \{1\}$. In this case, the series (I) and (II) are certainly identical and we have that $r = s = 0$. Thus the theorem holds trivially for all simple groups and in particular for all groups of orders less than four.

We shall proceed by induction on $|G|$ and we shall disregard simple groups, that is we shall henceforth assume that $r \geq 1$ and $s \geq 1$. Two cases have to be distinguished.

(i) $A_1 = B_1$. Omitting the first terms in (I) and (II) we obtain two composition series for A_1, namely

$$A_1 \rhd A_2 \rhd \ldots \rhd A_r \rhd \{1\}$$

and

$$A_1 \rhd B_2 \rhd \ldots \rhd B_s \rhd \{1\}.$$

Since $|A_1| < |G|$, the inductive hypothesis implies that the composition factors

$$A_1/A_2, \quad A_2/A_3, \quad \ldots, A_r$$

and

$$A_1/B_2, \quad B_2/B_3, \quad \ldots, B_s$$

are isomorphic in pairs. Since the first terms in $(I)'$ and $(II)'$ are now identical, we have that $(I) \sim (II)$, and the theorem is established in this case.

(ii) $A_1 \neq B_1$. Since $A_1 \lhd G$ and $B_1 \lhd G$, the group (see §15)

$$C = A_1 B_1 \quad (= B_1 A_1)$$

is normal in G and contains both A_1 and B_1; in particular,

$$G \geq C \geq A_1.$$

But A_1 is a maximal normal subgroup of G. Hence either $C = G$ or $C = A_1$. The latter alternative has to be rejected; for, since $B_1 < C$, it would imply that $G > A_1 > B_1$, which is incompatible with the fact that B_1 is maximal. Thus

$$G = A_1 B_1.$$

Let $D = A_1 \cap B_1$. On applying Theorem 10 (p. 73) we find that

$$G/A_1 \cong B_1/D, \quad G/B_1 \cong A_1/D. \qquad (6.5)$$

By construction G/A_1 and G/B_1 are simple groups, and hence so also are B_1/D and A_1/D, that is D is a maximal normal subgroup both of A_1 and of B_1. Let

$$D \rhd D_1 \rhd \ldots \rhd D_t \big| \rhd \{1\}$$

be any composition series for D. We can then construct two composition series for G, namely

$$G \rhd A_1 \rhd D \rhd D_1 \rhd \ldots \rhd D_t \rhd \{1\} \qquad (III)$$

and

$$G \rhd B_1 \rhd D \rhd D_1 \rhd \ldots \rhd D_t \rhd \{1\}. \qquad (IV)$$

Indeed all the composition factors

$$G/A_1, \quad A_1/D, \quad \vdots \quad D/D_1, \ldots, D_t \qquad (III)'$$

and

$$G/B_1, \quad B_1/D, \quad \vdots \quad D/D_1, \ldots, D_t \qquad (IV)'$$

are simple groups, as has just been observed. The composition factors on the right of the vertical line are common to $(III)'$ and $(IV)'$ whilst those on the left are isomorphic when arranged crosswise in pairs. Thus $(III) \sim (IV)$. Now (I) and (III) agree in the first two terms, a situation we discussed under (i). Hence $(I) \sim (III)$. Similarly, $(II) \sim (IV)$. Hence we conclude that $(I) \sim (II)$. This completes the proof.

We illustrate the theorem with two examples, which are rather trivial, because we have not yet met simple groups of composite order (see p. 139).

(1) Let G be the non-Abelian group of order 6 (p. 47). It can be defined by the relations

$$a^3 = b^2 = (ab)^2 = 1.$$

The subgroup $A = \mathrm{gp}\{a\}$ is of order 3 and of index 2 in G. Hence $A \lhd G$ (p. 62, (iv)), and

$$G \rhd A \rhd \{1\}$$

is a composition series, because the factors

$$G/A \cong C_2 \quad \text{and} \quad A \cong C_3 \tag{6.6}$$

are of prime order and therefore simple.

(2) Let $G = \mathrm{gp}\{s\}$ be the cyclic group of order 6. Then $A_2 = \mathrm{gp}\{s^2\}$ is a subgroup of order 3, and since all subgroups of an Abelian group are normal, we have the composition series

$$G \rhd A_2 \rhd \{1\}$$

with factors

$$G/A_2 \cong C_2 \quad \text{and} \quad A_2 \cong C_3. \tag{6.7}$$

Alternatively, we can start with the subgroup $A_3 = \mathrm{gp}\{s^3\}$ of order 2 and construct the composition series

$$G \rhd A_3 \rhd \{1\},$$

whose factors

$$G/A_3 \cong C_3 \quad \text{and} \quad A_3 \cong C_2$$

are the same as (6.7), but in the reverse order. We observe that the same composition factors occur in (1) and (2) although the groups are not isomorphic. In both cases the composition factors are of prime orders. This property characterizes a very important class of groups, which will be studied in the next section.

36. Soluble Groups

DEFINITION 7. *A finite group is said to be* **soluble** *if all its composition factors are of prime orders.*

The following proposition is often useful to decide whether a given group is soluble.

PROPOSITION 15. *The finite group G is soluble if it contains a normal subgroup H such that H and G/H are soluble.*

Proof. If these conditions are fulfilled we have composition series

$$H \rhd H_1 \rhd \ldots \rhd H_r \rhd \{1\} \tag{6.8}$$

and

$$G/H \rhd G_1/H \rhd \ldots \rhd G_s/H \rhd H. \tag{6.9}$$

(It should be remembered that any subgroup of G/H can be written in the form A/H and that H is the unit element of G/H.) By hypothesis, the composition factors of (6.8) and (6.9) are of prime orders, and in particular G_s/H has prime order. Since by Theorem 9 (p. 71)

$$\frac{G_{i-1}/H}{G_i/H} \cong G_{i-1}/G_i \quad (G_0 = G),$$

we infer that

$$G \rhd G_1 \rhd \ldots \rhd G_s \rhd H \rhd H_1 \rhd \ldots \rhd H_r \rhd \{1\}$$

is a composition series for G, in which each composition factor is of prime order. Hence G is soluble. The usefulness of this result is illustrated by the following applications.

PROPOSITION 16. *All finite Abelian groups are soluble.*

Proof. Let A be a finite Abelian group. If $|A| = p$, where p is a prime, then the composition series

$$A \rhd \{1\}$$

demonstrates the solubility of A. We now use induction on $|A|$ and assume that A is of composite order. Then A possesses a proper subgroup H, necessarily normal in A (exercise 4, Chapter II, p. 55). Since H and G/H are Abelian groups of orders less than $|A|$, the inductive hypothesis implies that H and G/H are soluble. Hence A is soluble by Proposition 15.

PROPOSITION 17. *All finite p-groups are soluble.*

Proof. Let P be a finite group such that $|P| = p^n$, where p is a prime. When $n = 1$, the group is certainly soluble; so we may use induction on n. By Theorem 7 (p. 59), the centre Z of P is non-trivial, and of course $Z \lhd P$. Now Z is soluble because it is Abelian, and P/Z is a p-group whose order is less than p^n. Hence P/Z is soluble by induction, and so also is P by virtue of Proposition 15.

Finally, we mention a characterization of soluble groups which appears to impose less stringent conditions than the original definition (p. 120).

PROPOSITION 18. *A finite group G is soluble if and only if it possesses subgroups $B_1, B_2, \ldots, B_s$ such that*

$$G \rhd B_1 \rhd B_2 \rhd \ldots \rhd B_s \rhd \{1\} \quad (G = B_0, \quad \{1\} = B_{s+1}) \quad (6.10)$$

E

and each

$$B_{i-1}/B_i \text{ is Abelian} \quad (i = 1, 2, \ldots, s+1). \tag{6.11}$$

Proof. If G is soluble, then by Definition 7 there exists a series (6.10) in which B_{i-1}/B_i is of prime order and therefore Abelian. Conversely, suppose that (6.10) and (6.11) hold. We may assume that there are no redundant terms in (6.10) so that each group is a proper subgroup of its predecessor. We proceed by induction on $|G|$. Ignoring the first term of (6.10) we obtain a series for B_1 which, by induction, implies that B_1 is soluble. Putting $i = 1$ in (6.11) we see that G/B_1 is Abelian and hence soluble. Hence, by Proposition 15, G is soluble.

37. The Derived Series. The derived group G' of G and some of its properties were introduced in §23. We recall (Theorem 11) that G' is the smallest normal subgroup having an Abelian quotient group. The process of forming the derived group can be iterated. Thus we construct the sequence

$$G \, (=G^0), \quad G', \quad G'' = (G')', \ldots, G^{(i)} = (G^{i-1})', \ldots.$$

Since $G^{(i)} \leq G^{(i-1)}$, we can write

$$G \geqq G' \geqq G'' \geqq \ldots \geqq G^{(i)} \geqq \ldots \tag{6.12}$$

This is called the **derived series** of G. Each group in (6.12) is not only normal in its predecessor but is a characteristic, and hence normal, subgroup of G itself (see example 14, Chapter III). The series may become stationary, that is $G^{(i+1)} = G^{(i)}$ for some i. Of course, this is bound to happen when G is finite. The most interesting case, however, is that in which (6.12) terminates in the unit group, as this yields yet another description of soluble groups.

THEOREM 20. *The finite group G is soluble if and only if its derived series terminates in the unit group, that is $G^{(s)} = \{1\}$ for some non-negative integer s.*

Proof. (i) Suppose that $G^{(s)} = \{1\}$, so that the derived series is

$$G > G' > \ldots > G^{(s-1)} > \{1\}. \tag{6.13}$$

By Theorem 11, $G^{(i-1)}/G^{(i)}$ is Abelian. Hence (6.13) is a series of the type considered in Proposition 18. It follows that G is soluble.

(ii) Assume that G is soluble and hence possesses a sequence of subgroups satisfying (6.10) and (6.11). We claim that

$$G^{(i)} \leq B_i \, (i = 1, 2, \ldots). \tag{6.14}$$

For since G/B_1 is Abelian, we deduce from Theorem 11 that $G' \leq B_1$. We now make the inductive hypothesis that $G^{(i-1)} \leq B_{i-1}$. From the very definition of a derived group it is clear that, whenever $K \leq L$, then $K' \leq L'$. Hence

$$G^{(i)} = (G^{(i-1)})' \leq B_{i-1}'.$$

Since B_{i-1}/B_i is Abelian, we invoke Theorem 11 once again to infer that $B_{i-1}' \leq B_i$, whence $G^{(i)} \leq B_i$. This establishes (6.14). When i is equal to $s+1$, this result becomes

$$G^{(s+1)} \leq B_{s+1} = \{1\}.$$

Thus the derived series terminates in the unit group.

38. Nilpotent groups. In this section we shall introduce a class of groups whose structure, next to those of Abelian groups, is most amenable to analysis. We begin by generalizing the notion of a commutator, which we defined in §23 (p. 74). Corresponding to any subsets A, B of G we can form the subgroup

$$[A, B] = \mathrm{gp}\{[a, b] \mid a \in A, b \in B\}. \tag{6.15}$$

Since

$$[a, b]^{-1} = (a^{-1}b^{-1}ab)^{-1} = b^{-1}a^{-1}ba = [b, a],$$

it follows that

$$[A, B] = [B, A], \tag{6.16}$$

because the inversion of each generator in (6.15) does not change the resulting group. It is obvious that, if $B \leq C$, then $[A, B] \leq [A, C]$.

With an arbitrary group G we associate a sequence of subgroups, inductively defined as follows:

$$\Gamma_1 = G, \quad \Gamma_2 = [G, G] = G', \ldots, \Gamma_{k+1} = [\Gamma_k, G]. \tag{6.17}$$

We shall show that $\Gamma_{k+1} \leq \Gamma_k \, (k = 1, 2, \ldots)$. This is trivial when $k = 1$. Assuming that $\Gamma_k \leq \Gamma_{k-1} \, (k > 1)$, we deduce that $\Gamma_{k+1} = [\Gamma_k, G] \leq [\Gamma_{k-1}, G] = \Gamma_k$. Thus (6.17) is in fact a descending series

$$G = \Gamma_1 \geq \Gamma_2 \geq \ldots \geq \Gamma_k \geq \Gamma_{k+1} \geq \ldots. \tag{6.18}$$

Each Γ_k is a characteristic subgroup of G, that is if α is an automorphism of G, then $\Gamma_k\alpha = \Gamma_k$ (see (3.50), p. 78). For, since $\alpha: G \rightarrow G$ is a homomorphism, we have that $[a, b]\alpha = [a\alpha, b\alpha]$ and hence $[A, B]\alpha = [A\alpha, B\alpha]$. Now $G\alpha = G$ and $\Gamma_{k+1}\alpha = [\Gamma_k\alpha, G]$. If we have already shown that $\Gamma_k\alpha = \Gamma_k$, which is trivial when $k = 1$, it follows that $\Gamma_{k+1}\alpha = [\Gamma_k, G] = \Gamma_{k+1}$. This proves that

$$\Gamma_k\alpha = \Gamma_k\,(k = 1, 2, 3, \ldots).$$

Consequently, $\Gamma_k \lhd G\,(k = 1, 2, 3, \ldots)$ and *a fortiori* $\Gamma_{k+1} \lhd \Gamma_k$. A less obvious property of (6.18) is expressed in the following proposition.

PROPOSITION 19. *The quotient group* Γ_k/Γ_{k+1} *lies in the centre of* $G/\Gamma_{k+1}\,(k = 1, 2, \ldots)$.

Proof: Let $v: G \rightarrow \Gamma_{k+1}$ be the natural map of G onto G/Γ_{k+1}, that is $xv = x\Gamma_{k+1} = \bar{x}$, say, where $x \in G$. The kernel of v is equal to Γ_{k+1}. A typical element of Γ_k/Γ_{k+1} is given by $\bar{u} = u\Gamma_{k+1}$, where u is an arbitrary element of Γ_k. We have to show that $\bar{u}$ and $\bar{x}$ commute for all x, that is we have to prove that $[\bar{u}, \bar{x}] = \bar{1}$, where $\bar{1}\,(=\Gamma_{k+1})$ is the unit element of G/Γ_{k+1}. Now

$$[\bar{u}, \bar{x}] = [uv, xv] = [u, x]v.$$

By the definition of Γ_{k+1}, $[u, x] \in \Gamma_{k+1}$. Hence $[u, x]v = \bar{1}$, which proves our assertion.

Next, we shall define an ascending series for an arbitrary group G. The construction is based on the following remark.

LEMMA. *Let U be a characteristic subgroup of G and let V/U be the centre of G/U. Then V is a characteristic subgroup of G.*

Proof. The group V may be described as the largest subgroup of G which commutes with G 'modulo U', that is

$$[V, G] \leqq U.$$

Indeed, if we apply the natural map $\mu: G \rightarrow G/U$, which 'kills' U, this relation becomes $[V\mu, G\mu] = \{\bar{1}\}$, where $\bar{1}$ is the unit element of G/U; this means that each element of $V/U\,(=V\mu)$ commutes with each element of $G/U\,(=G\mu)$. Suppose now that α is an automorphism of G. Then $[V\alpha, G] \leqq U\alpha$. By hypothesis, $U\alpha = U$ and therefore $[V\alpha, G] \leqq U$. Hence, by the maximality of V, $V\alpha \subset V$. If, instead of α, we use the automorphism α^{-1}, we deduce analogously that

$V\alpha^{-1} \subset V$, that is $V \subset V\alpha$. It follows that $V\alpha = V$. Thus V is a characteristic subgroup.

Now put $Z_0 = \{1\}$, and let Z_1 be the centre of G. Since Z_1 is a characteristic subgroup of G, we deduce from the lemma that there exists a characteristic subgroup Z_2 such that Z_2/Z_1 is the centre of G/Z_1. Proceeding inductively we define Z_{j+1} by the property that Z_{j+1}/Z_j is the centre of G/Z_j. Thus we construct an ascending series of characteristic subgroups

$$\{1\} = Z_0 \leqq Z_1 \leqq \ldots \leqq Z_j \leqq \ldots . \qquad (6.19)$$

As we remarked, the series (6.18) and (6.19) exist for any group G, but they may collapse into the first term if $G = G'$ ($= \Gamma_2$) or if $Z_1 = \{1\}$ respectively. We are more interested in the opposite case in which the series attain their maximal lengths by stretching from the group G itself down to the unit group $\{1\}$.

DEFINITION 8 (i). *A group G is said to have a* **lower central series** *of length r if*

$$G = \Gamma_1 > \Gamma_2 > \ldots > \Gamma_k > \ldots > \Gamma_r > \Gamma_{r+1} = \{1\}, \quad (6.20)$$

where $\Gamma_{k+1} = [\Gamma_k, G]$ $(k = 1, 2, \ldots, r)$.

(ii) *A group G is said to have an* **upper central series** *of length s if*
$$\{1\} = Z_0 < Z_1 < \ldots < Z_j < \ldots < Z_s = G, \qquad (6.21)$$

where Z_j/Z_{j-1} is the centre of G/Z_{j-1} $(j = 1, 2, \ldots, s)$.

As in the lemma, Z_j may be characterized as the largest subgroup of G having the property that

$$[Z_j, G] \leqq Z_{j-1}. \qquad (6.22)$$

There are some remarkable relationships between the terms of the two central series. In fact, we shall find that if one series exists, so does the other, and that both series have the same length.

First suppose that G possesses a lower central series of length r so that (6.20) holds, and consider the series (6.19) for this group. We claim that

$$\Gamma_{r+1-i} \leqq Z_i \quad (i = 0, 1, \ldots, r). \qquad (6.23)$$

This is evidently true when $i = 0$, because it is assumed that $\Gamma_{r+1} = \{1\} = Z_0$. Making the inductive hypothesis that (6.23) holds for a particular value of i we wish to prove that $\Gamma_{r-i} \leqq Z_{i+1}$. Since $\Gamma_{r+1-i} = [\Gamma_{r-i}, G]$, our hypothesis states that $[\Gamma_{r-i}, G] \leqq Z_i$.

By (6.22), Z_{i+1} is the greatest subgroup such that $[Z_{i+1}, G] \leqq Z_i$. It follows that $\Gamma_{r-i} \leqq Z_{i+1}$, thus proving (6.23) for all values of i. In particular, when $i = r$, we find that $\Gamma_1 = G \leqq Z_r$. This means that $Z_r = G$. Hence (6.19) terminates in G after at most r steps, that is G possesses an upper central series whose length, s, satisfies

$$s \leqq r. \tag{6.24}$$

Secondly, assume that (6.21) holds for G and examine the series (6.18) for this group. We now claim that

$$\Gamma_i \leqq Z_{s+1-i} \quad (i = 1, 2, \ldots, s+1). \tag{6.25}$$

This is true when $i = 1$, because we have assumed that $Z_s = G = \Gamma_1$. Proceeding by induction we suppose that (6.25) holds for a particular value of i, and we will show that $\Gamma_{i+1} \leqq Z_{s-i}$. Indeed, we have that $\Gamma_{i+1} = [\Gamma_i, G] \leqq [Z_{s+1-i}, G] \leqq Z_{s-i}$ as required. Putting $i = s+1$ in (6.25) we find that $\Gamma_{s+1} \leqq Z_0 = \{1\}$, that is $\Gamma_{s+1} = \{1\}$. Thus (6.18) terminates in $\{1\}$ after at most $s+1$ steps. This proves that G possesses a lower central series, whose length, r, satisfies $r \leqq s$; together with (6.24), this shows that $s = r$.

The foregoing investigation enables us at last to formulate the definition of the class of groups mentioned in the heading of this section.

DEFINITION 9. *A group G is said to be* **nilpotent** *if it possesses a lower central series or, equivalently, an upper central series. The common length of these series is called the nilpotency class of G.*

Example 1. If A is an Abelian group of order greater than one, then the upper central series reduces to

$$\{1\} = Z_0 < Z_1 = A$$

Thus the set of Abelian groups ($\neq \{1\}$) coincides with the set of nilpotent groups of class one.

Example 2. *Finite p-groups are nilpotent.* If P is a finite p-group, then by Theorem 7 (p. 59), its centre, Z_1, has order greater than one. Now P/Z_1 is also a p-group, and therefore its centre, Z_2/Z_1, is non-trivial, that is $Z_1 < Z_2$. Similarly P/Z_2 has a centre Z_3/Z_2, where $Z_2 < Z_3$. Continuing in this manner we construct a strictly ascending series

$$\{1\} = Z_0 < Z_1 < Z_2 < Z_3 < \ldots.$$

Since P is finite, this series must terminate. This happens, say, when $Z_r = P$. Thus P has an upper central series and is therefore nilpotent. From the many results about nilpotent groups we select one interesting fact, to which we shall refer again at the end of this book.

PROPOSITION 20. *If H is a proper subgroup of a nilpotent group G, then the normalizer N(H) of H in G is strictly greater than H.*

Proof. Let G be a nilpotent group of class r. It is trivial that $\{1\} = Z_0 \leqq H$. On the other hand, since H is a proper subgroup, $G = Z_r \nleqq H$. Hence there exists a unique integer k such that $0 \leqq k \leqq r-1$ and

$$Z_k \leqq H, \quad Z_{k+1} \nleqq H. \tag{6.26}$$

Thus there is an element u such that $u \in Z_{k+1}$ and $u \notin H$. It suffices to show that $u \in N(H)$, that is

$$u^{-1}Hu = H. \tag{6.27}$$

Let h_1 be an arbitrary element of H. Then

$$[u, h_1] \in [Z_{k+1}, G] = Z_k \leqq H$$

by (6.22) and (6.26). This means that $u^{-1}h_1^{-1}uh_1 = h_2$, where $h_2 \in H$. Hence $u^{-1}h_1^{-1}u \in H$. Since h_1^{-1} runs through H together with h_1, we have shown that $u^{-1}Hu \subset H$. Using the same argument when u is replaced by u^{-1}, we infer that $uHu^{-1} \subset H$, that is $H \subset u^{-1}Hu$. This establishes (6.27).

Exercises

(1) Find a composition series (i) for the dihedral group of order 8 (Table xi, p. 51) and (ii) for the quaternion group (Table xii, p. 51). Determine the composition factors in each case.

(2) Prove that every subgroup and quotient group of a soluble group is soluble.

(3) Verify that, if x, y, z are any elements of a group, (i) $[xy, z] = [x, z]^y[y, z]$; (ii) $[x, yz] = [x, z][x, y]^z$, where $a^t = t^{-1}at$.

(4) Show that if G is nilpotent of class 2, then G' lies in the centre of G and deduce the identities.
$$[xy, z] = [x, z][y, z], \quad [x, yz] = [x, z][x, y]$$
for such a group.

(5) Prove that every subgroup and factor group of a nilpotent group is nilpotent.

(6) Let G be nilpotent of class 3. Show that, if $v \in G'$ and $x \in G$, then $x^v = cx$, where $c \in Z$, the centre of G. Deduce that G' is Abelian.

(7) Prove that if M is a maximal subgroup of a nilpotent group G, then $M \lhd G$ and $|G/M| = p$, where p is a prime. [A maximal subgroup is a proper subgroup which is not contained in any other proper subgroup. Infinite groups need not possess maximal subgroups.]

(8) Let $D(2^n)$ be the dihedral group of order 2^{n+1} (Chapter II, example (7), p. 55), given by
$$a^{2^n} = b^2 = (ab)^2 = 1.$$
Prove that if Z_1 is the centre of $D(2^n)$, then $D(2^n)/Z_1 \cong D(2^{n-1})$. Deduce that $D(2^n)$ is nilpotent of class n.

VII. Permutation Groups

39. The Conjugacy Classes of S_n. In §7 (pp. 20–26) we introduced the family of symmetric groups S_n $(n = 1, 2, \ldots)$, and we described some of their elementary properties. The present chapter is devoted to a more detailed study of these groups, which together with their subgroups play a fundamental part in the theory of finite groups.

In this section we consider the problem of resolving S_n into its conjugacy classes (see §17). To this end we require a technique for computing the product $\tau^{-1}\alpha\tau$, where α and τ are arbitrary elements of S_n. Using the notation of (1.43) (p. 20) let

$$\alpha = \begin{pmatrix} 1 & 2 & \ldots n \\ a_1 & a_2 \ldots a_n \end{pmatrix}. \tag{7.1}$$

This symbol is abbreviated to

$$\alpha = \begin{pmatrix} i \\ a_i \end{pmatrix}, \tag{7.2}$$

where $i = 1, 2, \ldots, n$. In order to simplify the notation we let

$$\tau = \begin{pmatrix} 1 & 2 \ldots n \\ 1' & 2' \ldots n' \end{pmatrix} = \begin{pmatrix} i \\ i' \end{pmatrix} \tag{7.3}$$

where $1'2'\ldots n'$ is the rearrangement of $1\ 2\ldots n$ corresponding to τ. As we remarked on p. 20, the instructions contained in τ may equally well be presented in non-standard form, and in particular we may write

$$\tau = \begin{pmatrix} a_i \\ a_i' \end{pmatrix}, \tag{7.4}$$

where the first row of (7.4) is the same as the second row of (7.1). We now have that

$$\tau^{-1}\alpha\tau = \begin{pmatrix} i' \\ i \end{pmatrix}\begin{pmatrix} i \\ a_i \end{pmatrix}\begin{pmatrix} a_i \\ a_i' \end{pmatrix} = \begin{pmatrix} i' \\ a_i' \end{pmatrix}.$$

This result may be described as follows: in order to obtain $\tau^{-1}\alpha\tau$

operate with τ on each symbol in the expression for α, that is in both rows of (7.1), thus

$$\tau^{-1}\alpha\tau = \begin{pmatrix} i\tau \\ a_i\tau \end{pmatrix}. \tag{7.5}$$

Example. Let $n = 4$ and

$$\alpha = \begin{pmatrix} 1\ 2\ 3\ 4 \\ 2\ 4\ 1\ 3 \end{pmatrix}, \quad \tau = \begin{pmatrix} 1\ 2\ 3\ 4 \\ 1\ 4\ 2\ 3 \end{pmatrix}.$$

Operating with τ on each symbol of α we find that

$$\tau^{-1}\alpha\tau = \begin{pmatrix} 1\ 4\ 2\ 3 \\ 4\ 3\ 1\ 2 \end{pmatrix} = \begin{pmatrix} 1\ 2\ 3\ 4 \\ 4\ 1\ 2\ 3 \end{pmatrix}.$$

Next, apply this procedure to a cycle of degree m, say

$$\gamma = (a_1 a_2 \ldots a_m) = \begin{pmatrix} a_1 & a_2 \ldots a_{m-1} & a_m \\ a_2 & a_3 \ldots a_m & a_1 \end{pmatrix}.$$

Then by (7.5)

$$\tau^{-1}\gamma\tau = \begin{pmatrix} a_1' & a_2' \ldots a_{m-1}' & a_m' \\ a_2' & a_3' \ldots a_m' & a_1' \end{pmatrix} = (a_1' a_2' \ldots a_m'),$$

or more briefly

$$\tau^{-1}\gamma\tau = (a_1\tau\ a_2\tau \ldots a_m\tau). \tag{7.6}$$

We have seen (Theorem 2, p. 25) that every permutation α can be expressed as a product of disjoint cycles in an essentially unique manner, thus

$$\alpha = \gamma_1\gamma_2\ldots\gamma_r, \tag{7.7}$$

where $\gamma_1,\ \gamma_2,\ \ldots,\ \gamma_r$ are disjoint cycles involving

$$m_1, m_2, \ldots, m_r \tag{7.8}$$

objects respectively. For the present discussion it is convenient to retain cycles of unit length so that all n objects are listed in the product (7.7). The integers (7.8) are called the **cycle pattern** of α. It is convenient to arrange the numbers (7.8) according to increasing order of magnitude. Thus all possible cycle patterns of S_n are in one-to-one correspondence with the sets of integers (7.8) satisfying

$$1 \leqq m_1 \leqq m_2 \leqq \ldots \leqq m_r$$

and

$$m_1 + m_2 + \ldots + m_r = n, \tag{7.9}$$

r being arbitrary. Alternatively, if α contains e_1 cycles of degree 1, e_2 cycles of degree 2, ..., e_n cycles of degree n, the cycle pattern of α may be described by the non-negative integers

$$e_1, e_2, \ldots, e_n$$

satisfying

$$e_1 + 2e_2 + \ldots + ne_n = n. \tag{7.10}$$

The next result links the cycle patterns with the conjugacy classes of S_n.

PROPOSITION 21. *Two permutations are conjugate in S_n if and only if they have the same cycle pattern.*

Proof. Let α be resolved into disjoint cycles, thus

$$\alpha = \gamma_1\gamma_2\ldots\gamma_r = (x_1x_2\ldots)(y_1y_2\ldots)\ldots(w_1w_2\ldots),$$

where γ_i is of degree m_i and $m_1 + m_2 + \ldots + m_r = n$.

If τ is an arbitrary permutation, denoted as in (7.3), then

$$\beta = \tau^{-1}\alpha\tau = (\tau^{-1}\gamma_1\tau)(\tau^{-1}\gamma_2\tau)\ldots(\tau^{-1}\gamma_r\tau)$$

$$= (x_1'x_2', \ldots), (y_1'y_2'\ldots)\ldots(w_1'w_2'\ldots) = \gamma_1'\gamma_2'\ldots\gamma_r',$$

where $\gamma_1', \gamma_2', \ldots\gamma_r'$ are disjoint cycles, because τ is a one-to-one mapping. Hence β has the same cycle pattern as α.

Conversely, if α and β have the same cycle pattern, as above, the permutation

$$\tau = \begin{pmatrix} x_1 & x_2, \ldots, y_1 & y_2, \ldots, w_1 & w_2, \ldots \\ x_1' & x_2', \ldots, y_1' & y_2', \ldots, w_1' & w_2', \ldots \end{pmatrix}$$

has the property that $\tau^{-1}\alpha\tau = \beta$, so that α and β are conjugate.

Thus there are as many conjugacy classes in S_n as there are possible cycle patterns; expressed differently, the number, k, of conjugacy classes in S_n is equal to the number of partitions of n into positive summands (7.9) or into non-negative summands (7.10). When the latter interpretation is used, the partition is often denoted by

$$1^{e_1}2^{e_2}\ldots n^{e_n}, \tag{7.11}$$

components with zero frequencies usually being omitted in concrete cases; for example

$$1^3 \ 3 \ 4^2$$

is the partition $1+1+1+3+4+4$ of 14. Unfortunately, there is no simple formula which expresses the class number k as a function of n. The following table gives k for the first few values of n:

n	1	2	3	4	5.	6	7	8
k	1	2	3	5	7	11	15	22

[Table (xiii)l

For example, when $n = 5$, the partitions (7.11) are

$$1^5, \ 1^3\,2, \ 1^2\,3, \ 1\,2^2, \ 1\,4, \ 2\,3, \ 5.$$

On the other hand, it is not difficult to say how many elements lie in a particular conjugacy class of S_n.

PROPOSITION 22. (CAUCHY) *Suppose that α has a cycle pattern which corresponds to the partition $1^{e_1}2^{e_2}\ldots n^{e_n}$. Then the number of permutations which are conjugate with α in S_n is equal to*

$$h_\alpha = \frac{n!}{1^{e_1}e_1!\,2^{e_2}e_2!\ldots n^{e_n}e_n!}. \tag{7.12}$$

Proof: The cycle pattern of α may be indicated by the diagram

$$\underbrace{(.)(.)\ldots(.)}_{e_1} \ \underbrace{(..)(..)\ldots(..)}_{e_2} \ \ldots, \tag{7.13}$$

which corresponds to the partition (7.11).

There are precisely n spaces in (7.13), and we obtain an element of S_n by filling in the n objects in any manner. In each case we obtain a permutation which has the same cycle pattern as α. There are $n!$ ways of arranging the objects, but not all arrangements yield distinct elements of S_n. Consider the e_j cycles of degree j which appear in (7.13) ($1 \leq j \leq n$). In the first place, these e_j cycles can be permuted amongst themselves in $e_j!$ ways, without changing the resulting element of S_n; next, each cycle

$$(a_1a_2\ldots a_j)$$

can be written in j different ways, because

$$(a_1a_2\ldots a_j) = (a_2a_3\ldots a_ja_1) = \ldots = (a_ja_1\ldots a_{j-1}).$$

Thus each element of S_n has been counted $e_j!^{j^{e_j}}$ times in so far as cycles of degree j are concerned. Altogether, this particular element of the conjugacy class of α has been repeated $1^{e_1}e_1!\,2^{e_2}e_2!\ldots n^{e_n}e_n!$ times. Hence the number of distinct elements in the class is given by (7.12).

We know from Proposition 7 (p. 58) that h_α is the index of the centralizer of α in the group S_n. Thus we have the following result.

PROPOSITION 23. *If α is a permutation with cycle pattern* (7.11), *then the centralizer of α in S_n is of order*

$$1^{e_1}e_1! \, 2^{e_2}e_2! \ldots n^{e_n}e_n!. \tag{7.14}$$

Example. Let ϕ be a cycle which involves all n objects, say

$$\phi = (1 \ 2 \ldots n).$$

In this case, $e_1 = e_2 = \ldots = e_{n-1} = 0, e_n = 1$. Hence by (7.14) the centralizer of ϕ is of order n. But ϕ certainly commutes with $\iota(=\phi^0)$, ϕ, ϕ^2, ..., ϕ^{n-1}, which are n distinct elements of S_n. Thus in this case the centralizer of ϕ coincides with the cyclic group generated by ϕ.

40. Transpositions. A cycle of degree 2 is called a **transposition.** Thus a typical transposition, say

$$\tau = (ab) \tag{7.15}$$

interchanges a and b and leaves all other symbols fixed. We note that

$$\tau^2 = \iota, \quad \tau = \tau^{-1},$$

where ι is the identity permutation. The group S_n contains $\frac{1}{2}n(n-1)$ transpositions.

Next, we shall let S_n act on a set

$$x_1, x_2, \ldots, x_n \tag{7.16}$$

of indeterminates. Thus if α sends i into a_i we define

$$x_i\alpha = x_{a_i} \quad (i = 1, 2, \ldots, n),$$

and, more generally, if f is any function of the indeterminates (7.16), we put

$$f(x_1, x_2, \ldots, x_n)\alpha = f(x_{a_1}, x_{a_2}, \ldots, x_{a_n}). \tag{7.17}$$

In particular, we consider the **difference product**

$$\begin{aligned}
\Delta = \prod_{i<j} (x_i - x_j) = {} & (x_1 - x_2)(x_1 - x_3)(x_1 - x_4)\ldots(x_1 - x_n) \\
& \times (x_2 - x_3)(x_2 - x_4)\ldots(x_2 - x_n) \\
& \times (x_3 - x_4)\ldots(x_3 - x_n) \\
& \cdots \\
& \times (x_{n-1} - x_n). \tag{7.18}
\end{aligned}$$

Clearly, if the indeterminates are subjected to a permutation α, the function Δ either remains unchanged or else is multiplied by -1. Thus in the notation (7.17)

$$\Delta\alpha = \zeta(\alpha)\Delta, \qquad (7.19)$$

where $\zeta(\alpha) = \pm 1$.

DEFINITION 10. *The permutation α is said to be* **even** *or* **odd** *according as $\zeta(\alpha) = 1$ or $\zeta(\alpha) = -1$. The function $\zeta(\alpha)$ is called the* **alternating character** *of S_n.*

The most important fact about this function is expressed in the following proposition.

PROPOSITION 24. *If α and β are any permutations, then*

$$\zeta(\alpha\beta) = \zeta(\alpha)\zeta(\beta), \qquad (7.20)$$

that is the product of two even or two odd permutations is even, whilst the product of an even and an odd permutation is odd.

Proof. We apply the operation β to both sides of (7.19), and note that by the definition of operational composition,

$$(\Delta\alpha)\beta = \Delta(\alpha\beta).$$

Thus

$$\Delta(\alpha\beta) = \zeta(\alpha)\Delta\beta,$$

the constant $\zeta(\alpha)$ being unaffected by the action of β. Applying (7.19) to $\alpha\beta$ and to β we obtain the relation

$$\zeta(\alpha\beta)\Delta = \zeta(\alpha)\zeta(\beta)\Delta,$$

whence the assertion follows. More generally

$$\zeta(\alpha_1\alpha_2\ldots\alpha_r) = \zeta(\alpha_1)\zeta(\alpha_2)\ldots\zeta(\alpha_r). \qquad (7.21)$$

The definition of $\zeta(\alpha)$ may be recast in such a way that the function Δ no longer appears explicitly. Each factor $(x_i - x_j)$ of Δ corresponds to a pair (i, j) of integers, such that $1 \leq i < j \leq n$. After the application of α, which sends i into α_i and j into α_j, this factor becomes $(x_{\alpha_i} - x_{\alpha_j})$. This is a factor of Δ if $\alpha_i < \alpha_j$, whilst Δ contains the factor $-(x_{\alpha_i} - x_{\alpha_j})$ if $\alpha_i > \alpha_j$. We say, the pair (i, j) causes an **inversion** if $i-j$ and $\alpha_i - \alpha_j$ are of opposite sign. Let t be the total number of inversions when all pairs (i, j) are considered. Then

$$\zeta(\alpha) = (-1)^t.$$

The number t is readily found as follows: write the permutation α in standard form, for example

$$\alpha = \begin{pmatrix} 1 & 2 & 3 & 4 & 5 & 6 \\ 3 & 4 & 6 & 2 & 5 & 1 \end{pmatrix}$$
$$\| \quad \| \quad \|\| \quad | \quad | \qquad (t = 9)$$

Let k be any number in the second row. If k is followed by $s(\geqq 0)$ integers less than k, we say that k has a *score* of s. The score for each k is recorded, whence the total score, t, is readily found. For example, 3 has the score two, because it is followed by 2 and 1, and 6 has the score three, because it is followed by 2, 5 and 1. In the present example $t = 9$, and so $\zeta(\alpha) = -1$.

Clearly, the identity permutation, ι, leaves Δ unchanged so that

$$\zeta(\iota) = 1. \tag{7.22}$$

Next, for any permutation α,

$$\zeta(\alpha)\zeta(\alpha^{-1}) = \zeta(\iota) = 1,$$

whence

$$\zeta(\alpha) = \zeta(\alpha^{-1}), \tag{7.23}$$

that is, inverse permutations have the same character.

If α and β are arbitrary permutations

$$\zeta(\beta^{-1}\alpha\beta) = \zeta(\beta^{-1})\zeta(\alpha)\zeta(\beta) = \zeta(\alpha).$$

Thus conjugate permutations have the same character, that is ζ is constant on each conjugacy class of S_n.

Let τ be a transposition, as in (7.15). Then by Proposition 21, τ is conjugate with the particular transposition $\sigma = (12)$. The action of σ changes the sign of $x_1 - x_2$ and replaces the remaining factors in the first row of (7.18) by those of the second row without introducing further minus signs. Hence $\zeta(\sigma) = -1$ and therefore $\zeta(\tau) = -1$. Thus all transpositions are odd permutations.

In order to find the character of a cycle of degree m we use the formula

$$(a_1 a_2 \ldots a_m) = (a_1 a_2)(a_1 a_3) \ldots (a_1 a_m), \tag{7.24}$$

which is easily verified by evaluating the product on the right:

$$a_1 \rightarrow a_2, a_2 \rightarrow a_1 \rightarrow a_3, a_3 \rightarrow a_1 \rightarrow a_4, \text{ and so on.}$$

Since $m-1$ transposition factors are involved we obtain that

$$\zeta(a_1 a_2 \ldots a_m) = (-1)^{m-1}. \tag{7.25}$$

The following consequence of (7.24) is worth recording.

THEOREM 21. *Every permutation can, in many ways, be expressed as the product of transpositions. The number of transposition factors in any such product is either always even or always odd, according as the given permutation is even or odd.*

Proof. Let α be the given permutation. We already know (p. 25) that α can be expressed as a product of cycles. By (7.24) each cycle is a product of transpositions. Thus we certainly have that

$$\alpha = \tau_1 \tau_2 \ldots \tau_s, \tag{7.26}$$

where each τ is a transposition. The product is not unique; for example we can insert pairs of factors

$$(ab)(ab),$$

which are equivalent to the identity permutation. Less trivially, if $a \neq 1$ and $b \neq 1$, we have the relation

$$(ab) = (1a)(1b)(1a), \tag{7.27}$$

and there are analogous relations in which the object 1 is replaced by any object distinct from a and b. However, (7.26) implies that $\zeta(\alpha) = (-1)^s$, and since $\zeta(\alpha)$ is determined by α alone, it follows that s is even or odd according as α is even or odd.

Using the nomenclature introduced in §12 (p. 38) we have the

COROLLARY. *The group S_n is generated by the set of transpositions. By virtue of (7.27) this result may be made more precise.*

PROPOSITION 25. *The group S_n is generated by the $n-1$ transpositions*

$$(1\ 2), (1\ 3), \ldots, (1\ n).$$

41. The Alternating Group. We return to the distinction between the even and odd permutations introduced in Definition 10 (p. 134), and we begin with a simple result about an arbitrary permutation group, that is any subgroup of S_n for a suitable value of n.

PROPOSITION 26. *In every permutation group G the even permutations form a normal subgroup which is either equal to G or else is of index two in G.*

Proof. Let H be the set of even permutations in G. By (7.20), (7.22) and (7.23), H is a subgroup G. If $H = G$, we have nothing more to prove. If $H \neq G$, then G contains at least one odd permutation σ, and the coset $H\sigma$ is distinct from H. Let δ be an arbitrary odd permutation of G. Then $\sigma\delta^{-1}$ is even, that is $\sigma\delta^{-1} \in H$ and hence $H\sigma = H\delta$ (Proposition 5, p. 33). Thus there are precisely two cosets of H in G, so that $[G:H] = 2$, as asserted. By remark (iv) on p. 62, H is normal in G.

Special interest is attached to the case in which $G = S_n$.

DEFINITION 11. *The set of all even permutations of S_n ($n \geq 2$), forms a group A_n of order $\frac{1}{2}n!$, which is called the* **alternating** *group of degree n.*

For example, the group A_4 is of order $\frac{1}{2}(4!) = 12$ and consists of the following permutations (arranged according to conjugacy classes of S_4).

$$A_1 = C_0 \cup C_1 \cup C_2,$$

where

$$C_0 = \iota$$
$$C_1 = (12)(34) \cup (13)(24) \cup (14)(23) \tag{7.28}$$
$$C_2 = (123) \cup (124) \cup (132) \cup (134) \cup (142) \cup (143) \cup (234) \cup (243).$$

It may be asked whether S_n possesses any other proper normal subgroups besides A_n. Ignoring the trivial cases in which $n = 1$ or $n = 2$, we shall answer this question from first principles when $n = 3$ or $n = 4$ by using the remark (p. 62, (iii)) that a normal subgroup must be the union of complete conjugacy classes, including the class consisting of the unit element.

The classes of S_3 are

$$\iota, \quad (12) \cup (13) \cup (23) \quad \text{and} \quad (123) \cup (132)$$

containing 1, 3 and 2 elements respectively. It is only when we join the unit element to the last class that we get a set in which the number of elements divides $|S_3|(=6)$, as is necessary for a subgroup. Indeed,

$$A_3 = \iota \cup (123) \cup (132),$$

and this is therefore the only proper normal subgroup of S_3.

The group S_4 has five conjugacy classes (see table (xiii), p. 132). Three of these classes, which consist of even permutations, are listed in (7.28), the remaining two classes are

$$C_3 = (12) \cup (13) \cup (14) \cup (23) \cup (24) \cup (34) \quad \text{and}$$

$$C_4 = (1234) \cup (1243) \cup (1324) \cup (1342) \cup (1423) \cup (1432).$$

Since $|C_0| = 1, |C_1| = 3, |C_2| = 8, |C_3| = |C_1| = 6$, only

$$V = C_0 \cup C_1 \quad \text{and} \quad A_4 = C_0 \cup C_1 \cup C_2$$

have cardinals dividing $|S_4| \ (=24)$, as is required for subgroups. We already know that $A_4 \lhd S_4$, and it is a remarkable fact that

$$V = \iota \cup (12)(34) \cup (13)(24) \cup (14)(23)$$

happens to be a group. For if we put $\alpha = (12)(34)$ and $\beta = (13)(24)$, then $\alpha\beta = \beta\alpha = (14)(23)$ and $\alpha^2 = \beta^2 = \iota$. Thus $V \lhd S_4$, and V possesses the structure of the four-group (p. 46). We have now proved that A_4 and V are the only proper normal subgroups of S_4. Incidentally, since V consists only of even permutations, we have that $V \lhd A_4$.

In the composition series (§35)

$$S_3 \rhd A_3 \rhd \{\iota\}$$

$$S_4 \rhd A_4 \rhd V \rhd \{\iota\}$$

all composition factors are of prime orders. This proves that S_3 and S_4 are soluble groups (§36). We shall see later on that in this respect S_n behaves differently when $n > 4$.

It is useful to have at our disposal a fairly straightforward set of generators for the group A_n.

PROPOSITION 27. When $n \geq 3$, the group A_n can be generated by the $n-2$ ternary cycles

$$(123), (124) \ldots, (12n). \tag{7.29}$$

Proof. By Proposition 25, every permutation can be expressed as a product of transpositions of the type $(1i)$. For an even permutation the number of transposition factors must be even. Hence A_n is generated by pairs of factors $(1\ i)(1\ j)$. Since $(1\ i)^2 = \iota$, we may suppose that in each pair $i \neq j$. Now

$$(1\ i)(1\ j) = (1\ i\ j). \tag{7.30}$$

It $i = 2$, the pair of transpositions is equal to one of the ternary cycles listed in (7.29). If $j = 2$, we observe that

$$(1\ i)(1\ 2) = (1\ i\ 2) = (1\ 2\ i)^2.$$

Finally, if $i > 2$ and $j > 2$, we use the relation

$$(1\ i\ j) = (1\ 2\ j)\ (1\ 2\ i)\ (1\ 2\ j)^{-1}.$$

Thus in all cases the right-hand side of (7.30) can be expressed in terms of the generators (7.29).

Recalling the notion of a simple group (p. 61) we shall now establish a celebrated result about alternating groups, which is due to E. Galois.

THEOREM 22. *When $n \neq 4$, the group A_n is simple.*

Proof. We have seen (p. 61) that V is a proper normal subgroup of A_4. Hence A_4 is not simple. From now on we shall assume that $n > 4$. The theorem is equivalent to the following statement: if $N \lhd A_n$ and $|N| > 1$, then $N = A_n$. The crucial hypothesis is that N is normal in A_n. Thus if $\alpha \in N$ and if δ is an arbitrary even permutation then $\delta^{-1}\alpha\delta \in N$, and hence also $\delta^{-1}\alpha\delta\alpha^{-1} \in N$. The proof of this theorem is broken down into several steps.

(i) Suppose that N contains a 3-cycle, say

$$\alpha = (a\ b\ c).$$

We shall then prove that N contains all 3-cycles

$$\xi = (x\ y\ z),$$

where x, y, z are any preassigned distinct objects. By Proposition 27 this immediately implies that $N = A_n$.

The permutation

$$\phi = \begin{pmatrix} a & b & c \\ x & y & z \end{pmatrix}$$

is an element of S_n with the understanding that any object not mentioned in ϕ remains fixed. By (7.7) we have that

$$\phi^{-1}\alpha\phi = \xi.$$

Since $n \geq 5$, there are at least two objects, e, f which are not included in α. The transposition $\tau = (e\ f)$ commutes with α, and it follows that

$$(\tau\phi)^{-1}\alpha(\tau\phi) = \xi.$$

Plainly, either ϕ or $\tau\phi$ belongs to A_n. Hence α is conjugate with ξ in A_n, and we conclude that $\xi \in A_n$.

(ii) Next, we assume that N contains the permutation

$$\omega = \gamma\delta\varepsilon\dots, \tag{7.31}$$

where γ, δ, ε, $\dots$ are disjoint cycles and the degree of γ exceeds three, say

$$\gamma = (a_1a_2a_3a_4, \dots, a_m), \quad m > 3.$$

Now $\sigma = (a_1a_2a_3)$ is an even permutation which commutes with all cycles of (7.31) except the first. Thus

$$\omega_1 = \sigma^{-1}\omega\sigma = (\sigma^{-1}\gamma\sigma)\delta\varepsilon\dots$$

belongs to N, and so does $\omega_1\omega^{-1}$. Since δ, ε, $\dots$ commute with both γ and $\sigma^{-1}\gamma\sigma$, we find that

$$\begin{aligned}
\omega_1\omega^{-1} &= \sigma^{-1}\gamma\sigma\gamma^{-1} \\
&= (a_2a_3a_1a_4\dots a_m)(a_ma_{m-1}\dots a_4a_3a_2a_1) \\
&= (a_1a_3a_m).
\end{aligned}$$

Hence N contains a 3-cycle, and we deduce from (i) that $N = A_n$. From now on we may suppose that all permutations of N are products of disjoint cycles whose degrees are 1, 2 or 3.

(iii) Suppose N contains a permutation ω which involves at least two 3-cycles, say

$$\omega = \alpha\beta\lambda,$$

where $\alpha = (a_1a_2a_3)$, $\beta = (b_1b_2b_3)$, and λ does not depend on a_i or b_i $(i = 1, 2, 3)$. Choosing

$$\sigma = (a_2a_3b_1)$$

we observe that σ commutes with λ. Hence N contains the element

$$\begin{aligned}
\sigma^{-1}\omega\sigma\omega^{-1} &= (\sigma^{-1}\alpha\sigma)(\sigma^{-1}\beta\sigma)\alpha^{-1}\beta^{-1}) \\
&= (a_1a_3b_1)(a_2b_2b_3)(a_3a_2a_1)(b_3b_2b_1) \\
&= (a_1a_2b_1a_3b_3),
\end{aligned}$$

which contradicts the hypothesis that no cycles of degree greater than three occur in N.

(iv) When only a single 3-cycle is permitted amongst the factors, a typical element is of the form

$$\omega = (a_1a_2a_3)\lambda,$$

where λ is a product of disjoint transpositions. Thus $\lambda^2 = \iota$, and N contains the element

$$\omega^2 = (a_1 a_3 a_2),$$

which brings us back to (i).

(v) Finally, we must discuss the case in which all elements of N, other than ι, are products of disjoint transpositions. When $n = 4$, this situation actually occurs and leads the group V mentioned on p. 138. However, since we are assuming that $n > 4$, we may argue as follows: as the number of transposition factors must be even, a typical element of N is of the form

$$\omega = (a_1 a_2)(b_1 b_2)\lambda,$$

where λ does not involve a_1, a_2, b_1, b_2. Choosing a fifth element c distinct from those just mentioned we use in turn the transforming elements $\sigma = (a_2 b_1 b_2)$ and $\delta = (a_1 b_2 c)$ to construct from ω further elements of N as follows:

$$\omega_1 = \sigma^{-1}\omega\sigma = (a_1 b_1)(b_2 a_2)\lambda,$$

$$\omega_2 = \omega_1\omega^{-1} = (a_1 b_1)(b_2 a_2)(a_1 a_2)(b_1 b_2)$$

$$= (a_1 b_2)(a_2 b_1),$$

$$\omega_3 = \delta^{-1}\omega_2\delta = (b_2 c)(a_2 b_1),$$

$$\omega_3\omega_2^{-1} = (b_2 c)(a_2 b_1)(a_1 b_2)(a_2 b_1) = (a_1 b_2 c),$$

Thus, contrary to our hypothesis, N would contain a 3-cycle after all. This concludes the proof of the theorem.

We can now revert to the question regarding the normal subgroups of S_n when $n > 4$.

PROPOSITION 28. *When $n > 4$, the only proper normal subgroup of S_n is the alternating group A_n.*

Proof: Suppose that $H \lhd S_n$ and $|H| > 1$. First, we shall show that H cannot be of order 2. For suppose that

$$H = \{\iota, \xi \mid \xi^2 = \iota\}.$$

Then ξ must either be a transposition or a product of disjoint transpositions. In the former case, let $\xi = (ab)$. There exists an object c, distinct from a and b. Since $H \lhd S_n$, the element $(ac)^{-1}(ab)(ac) = (bc)$ would belong to H, and H would contain more than two elements. Next, suppose that $\xi = (a_1a_2)(b_1b_2)\lambda$, where λ is independent of a_1, a_2, b_1, b_2. Then, if $\sigma = (a_2b_1b_2)$, $\sigma^{-1}\xi\sigma \in H$, but $\sigma^{-1}\xi\sigma \neq \xi$, which contradicts the assumption that $|H| = 2$. Thus $|H| > 2$. By Proposition 26, at least one half of the elements of H are even; hence, if $D = H \cap A_n$, then $|D| > 1$. Evidently, $D \lhd A_n$. Since A_n is simple, $D = A_n$, which means that

$$A_n \leq H. \tag{7.32}$$

As H is a proper subgroup of S_n, we have that $|H| \leq \frac{1}{2}n!$ Hence $|A_n| = |H|$, and we infer from (7.32) that $A_n = H$.

42. Permutation Representations. It was not until the beginning of the twentieth century that the abstract concept of a group was fully appreciated and accepted by mathematicians. The earlier literature on the subject, including the classical works of Cauchy, Galois and C. Jordan, dealt almost exclusively with permutation groups, that is with subgroups of the symmetric groups S_n. However, many of their results applied equally well to arbitrary finite groups and were independent of the assumption that the elements of the group are permutations. Even in the context of modern group theory the study of permutation groups is a topic of great interest. Not only do these groups provide a wealth of fairly accessible examples of finite groups, but, as A. Cayley observed in 1854, every finite group is isomorphic with a permutation group.

Let

$$G: a_1, a_2, \ldots, a_g \tag{7.33}$$

be a finite of order g. If x is any one of these elements, the products

$$a_1x, a_2x, \ldots, a_gx \tag{7.34}$$

are g distinct elements of G and therefore constitute the whole group. Thus (7.34) is a rearrangement of (7.33), that is we can associate with x the permutation

$$x\rho = \begin{pmatrix} a_1 & a_2 & ..a_g \\ a_1x & a_2x..a_gx \end{pmatrix}$$

of degree g. The objects on which this permutation acts, are the group elements themselves. It is convenient to use the abbreviated notation

$$x\rho = \begin{pmatrix} a_i \\ a_ix \end{pmatrix} \quad (i = 1, 2, \ldots, g). \tag{7.35}$$

The action of $x\rho$ on G may be briefly described by saying that each element of G is multiplied on the right by x. The order in which the elements are listed is immaterial. In particular, if u is a fixed element of G, the products $a_i u$ ($i = 1, 2, \ldots, g$) are all the elements of G, as we have remarked in (7.34). Therefore we can write

$$x\rho = \begin{pmatrix} a_i u \\ a_i u x \end{pmatrix} \tag{7.36}$$

Now let y be another element of G and let

$$y\rho = \begin{pmatrix} a_i \\ a_i y \end{pmatrix} \tag{7.37}$$

be the permutation associated with y. Computing the product of the permutations (7.35) and (7.37) we find that, by virtue of (7.36),

$$(x\rho)(y\rho) = \begin{pmatrix} a_i \\ a_i x \end{pmatrix}\begin{pmatrix} a_i \\ a_i y \end{pmatrix} = \begin{pmatrix} a_i \\ a_i x \end{pmatrix}\begin{pmatrix} a_i x \\ a_i xy \end{pmatrix} = \begin{pmatrix} a_i \\ a_i xy \end{pmatrix}.$$

Thus

$$(x\rho)(y\rho) = (xy)\rho, \tag{7.38}$$

which shows that the map

$$\rho : G \to S_g$$

is a homomorphism of G into S_g. Moreover, ρ is injective, that is its kernel consists of the identity element 1 of G alone (Proposition 9, p. 67). For suppose that

$$x\rho = \iota,$$

the identity element of S_g. This means that

$$a_i x = a_i \quad (i = 1, 2, \ldots, g),$$

which evidently implies that $x = 1$. Indeed, if $x \neq 1$, $x\rho$ displaces each element of G. Since ρ is injective, the image of G under ρ is isomorphic with a subgroup of S_g.

Next, we shall resolve $x\rho$ into disjoint cycles. Let x be of order r, thus

$$x^r = 1. \tag{7.39}$$

Beginning with any element a of G, we know that the action of $x\rho$ changes a into ax, which in turn is changed into ax^2; the image of

ax^2 is ax^3, and so on until we come to ax^{r-1}, whose image, by (7.39), is equal to a. Hence $x\rho$ contains the cycle

$$(a, ax, ax^2, \ldots, ax^{r-1}), \tag{7.40}$$

which involves r distinct elements of G. If $r < g$, we choose an element b not contained in (7.40), and we can construct a further cycle

$$(b, bx, bx^2, \ldots, bx^{r-1}). \tag{7.41}$$

Clearly (7.40) and (7.41) have no element in common; for, if they had, it would follow that $b = ax^t$ ($0 \le t \le r-1$), contrary to the choice of b. Continuing in this way we build up cycles, each containing r elements, until all g elements of G are accounted for. Thus

$$x\rho = (a, ax, \ldots, ax^{r-1})(b, bx, \ldots, bx^{r-1})\ldots(f, fx, \ldots, fx^{r-1}),$$

say. A permutation, in which all constituent cycles have the same length, is called a **regular permutation.** Incidentally, the last formula confirms that r is a factor of g.

We summarize our results as follows:

THEOREM 23. (CAYLEY). *Let $G: a_1, a_2, \ldots, a_g$ be an abstract group of order g. With each element x of G we associate the regular permutation*

$$x\rho = \begin{pmatrix} a_1 & a_2 \ldots a_g \\ a_1 x & a_2 x \ldots a_g x \end{pmatrix}.$$

The map $\rho: G \to S_g$ defined in this way is an injective homomorphism, so that G is isomorphic with a subgroup of S_g. If x is of order r, then $x\rho$ is the product of g/r cycles of degree r.

When an abstract group G is isomorphic with a group G' whose elements are concrete mathematical entities, such as permutations or matrices, we say that G' is a **faithful representation** of G in terms of permutations or matrices, as the case may be. All properties of G are also possessed by G'. Conversely, any information about G', which does not depend on the special nature of its elements, equally applies to G. Since it is often more convenient to carry out computations with concrete elements, the existence of a representation may enable us to elucidate the structure of an abstract group. The procedure is analogous to the use of coordinates in the discussion of geometrical problems. The particular representation furnished by Cayley's Theorem is known as the **right regular representation** of G. When G

is given by its multiplication table (§4, p. 11), the right regular representation can be read off immediately: in the two-line symbol for $x\rho$ the top line is the column headed by 1, and the bottom line is the column headed by x; indeed, a knowledge of the right regular representation is virtually equivalent to the construction of the multiplication table.

Example. In the case of the non-Abelian group of order 6, given in table (v), p. 12, the elements of the right regular representation resolved into cycles, are as follows:

$$1\rho = \begin{pmatrix} 1 & a & b & c & d & e \\ 1 & a & b & c & d & e \end{pmatrix} = (1)\ (a)\ (b)\ (c)\ (d)\ (e)$$

$$a\rho = \begin{pmatrix} 1 & a & b & c & d & e \\ a & b & 1 & d & e & c \end{pmatrix} = (1\ a\ b)\ (c\ d\ e)$$

$$b\rho = \begin{pmatrix} 1 & a & b & c & d & e \\ b & 1 & a & e & c & d \end{pmatrix} = (1\ b\ a)\ (c\ e\ d)$$

$$c\rho = \begin{pmatrix} 1 & a & b & c & d & e \\ c & e & d & 1 & b & a \end{pmatrix} = (1\ c)\ (a\ e)\ (b\ d)$$

$$d\rho = \begin{pmatrix} 1 & a & b & c & d & e \\ d & c & e & a & 1 & b \end{pmatrix} = (1\ d)\ (a\ c)\ (b\ e)$$

$$e\rho = \begin{pmatrix} 1 & a & b & c & d & e \\ e & d & c & b & a & 1 \end{pmatrix} = (1\ e)\ (a\ d)\ (b\ c).$$

It is sometimes more convenient to write ρ_x instead of $x\rho$ for a typical element of the right regular representation. Thus ρ_x may be briefly described by the formula

$$a\rho_x = ax\ (a \in G). \tag{7.42}$$

More generally, we may consider homomorphisms

$$\theta : G \to S_n,$$

not necessarily injective (faithful), where n is a suitable integer. When such a homomorphism exists, we say that G has a **permutation representation** of degree n. A fairly general method of constructing such representations is as follows: let H be a subgroup of G and let

$$G = Ht_1 \cup Ht_2 \cup \ldots \cup Ht_n \tag{7.43}$$

be the decomposition of G into right cosets relative to H, where $n = [G:H]$ (see p. 33). If x is a fixed element of G, the right cosets $Ht_i x$ $(i = 1, 2, \ldots, n)$ are distinct and must therefore be the same as those listed in (7.43). Thus

$$x\theta = \begin{pmatrix} Ht_1 & Ht_3 & \ldots Ht_n \\ Ht_1 x & Ht_2 x \ldots Ht_n x \end{pmatrix}$$

is a permutation of degree n, the objects being the n right cosets of H in G. By an argument similar to that used on p. 143, the map θ can easily be shown to be a homomorphism, that is

$$(x\theta)(y\theta) = (xy)\theta.$$

If k lies in the kernel of θ, we must have that

$$Ht_i k = Ht_i \ (i = 1, 2, \ldots, n),$$

which is equivalent to the condition that $t_i k \in Ht_i$ or $k \in t_i^{-1} Ht_i$ $(i = 1, 2, \ldots, n)$. Now any subgroup conjugate with H is of the form $t_i^{-1} Ht_i$ for a suitable value of i. For if y is any element of G, it lies in one of the cosets, say $y \in Ht_i$, that is, $y = ut_i$ where $u \in H$. Hence $y^{-1} Hy = t_i^{-1} u^{-1} Hut_i = t_i^{-1} Ht_i$. Thus we may say that the kernel of θ consists of the intersection of all groups conjugate with H. We collect these results in the following theorem.

THEOREM 24. *Let H be a subgroup of G of finite index n, and let $t_1, t_2, \ldots, t_n$ be a transversal (p. 33) of H in G. With each element x of G we associate the permutation*

$$x\theta = \begin{pmatrix} Ht_1, & Ht_2, & \ldots, Ht_n \\ Ht_1 x, & Ht_2 x, & \ldots, Ht_n x \end{pmatrix}.$$

The map $\theta: G \to S_n$ defined in this way is a homomorphism. The kernel of θ consists of the intersection of all the groups conjugate with H.

We conclude this section with an example that illustrates how these ideas can be used to obtain information about the structure of a group.

Example: The alternating group A_5 has no subgroups of orders 30 or 20 or 15. In other words, we claim that, if H is a proper subgroup of A_5, then $[A_5 : H] \geqq 5$. Suppose that H is a proper subgroup and put $[A_5 : H] = n$. By Theorem 24, there exists a homomorphism $\theta: A_5 \to S_n$. Let K be the kernel of θ. We know (p. 67) that K is a normal subgroup of A_5. But A_5 is a simple group (Theorem 22). Hence either $K = \{1\}$ or $K = A_5$. The second alternative can be discarded at once, for by Theorem 24, K is contained in H and therefore $|K| \leq |H| < |A_5|$. We must therefore have that $K = \{1\}$, that is θ is injective. Thus the image of A_5 under θ consists of 60 distinct elements of S_n, and this is impossible unless $n \geq 5$.

43. Transitive Groups. In this and the following section we consider permutations of a fixed degree, that is we are concerned with subgroups G of a particular symmetric group S_n. The objects on which G acts will be denoted by $1, 2, \ldots, n$ or by letters $a, b, \ldots$.

DEFINITION 12. *A group of permutations is said to be* **transitive** *if, given any pair of letters a, b (which need not be distinct), there exists at least one permutation in the group which transforms a into b. Otherwise the group is called* **intransitive**.

It should be observed that this concept applies only to permutation groups.

A permutation which changes a into b will be denoted by θ_{ab}, irrespective of its effect on the other symbols. There may, of course, be many such permutations for a given pair a, b. We note that θ_{ab}^{-1} changes b into a.

Evidently, the symmetric group S_n is transitive, as it contains all possible permutations, including the transposition (a, b), which can serve as θ_{ab}.

On the other hand, the group

$$V_1:(1), \quad (12), \quad (34), \quad (12)(34)$$

of order and degree 4 is intransitive because none of its permutations changes 1 into 3. Incidentally, this group is isomorphic with the group

$$V_2:(1), \quad (12)(34), \quad (13)(24), \quad (14)(23),$$

which, on the contrary, is transitive. Both groups are isomorphic with the four-group (table (iii), p. 12).

The set of permutations of G which leave the symbol 1 unaltered forms a subgroup G_1; for the identical permutation certainly belongs to this set, as does the inverse of any of its members and the product of any two of them. We call G the **stabilizer** of 1. The stabilizer G_a of the object a is defined analogously.

THEOREM 25. *A permutation group G of degree n is transitive if and only if the stabilizer G_1 is of index n in G.*

Proof. (i) Suppose G is transitive. By hypothesis G contains permutations

$$\theta_{11}, \theta_{12}, \ldots, \theta_{1n}, \tag{7.44}$$

which transform 1 into 1, 2, ..., n respectively. The right cosets

$$G_1\theta_{11}, G_1\theta_{12}, \ldots, G_1\theta_{1n} \tag{7.45}$$

are distinct, because all members of $G_1\theta_{1i}$ transform 1 into i and therefore differ from those of $G_1\theta_{1j}$, when $i \neq j$. It remains to show that (7.45) is a complete list of cosets. Let ξ be any element of G and suppose that ξ transforms 1 into a. Then $\xi\theta_{1a}^{-1}$ leaves 1 unchanged, that is $\xi\theta_{1a}^{-1} \in G_1$. Hence $\xi \in G_1\theta_{1a}$, which proves that the union of the cosets (7.45) is the whole group. Thus $[G:G_1] = n$.

(ii) Conversely, assume that G_1 is of index n and let

$$G = G_1\tau_1 \cup G_1\tau_2 \cup \ldots \cup G_1\tau_n$$

be a coset decomposition of G relative to G_1. First, we observe that no two of the permutations

$$\tau_1, \tau_2, \ldots, \tau_n \tag{7.46}$$

have the same effect on the object 1. For suppose that both τ_i and τ_j transform 1 into a. Then $\tau_i\tau_j^{-1}$ leaves 1 fixed, so that $\tau_i\tau_j^{-1} \in G_1$ and therefore $G_1\tau_i = G_1\tau_j$ (Proposition 5, p. 33), which is impossible unless $i = j$. Hence the permutations (7.46) can, in some order, be taken for the permutations (7.44). It is convenient to arrange (7.46) in such a way that $\theta_{1i} = \tau_i$ $(i = 1, 2, \ldots, n)$. Finally, if a, b is any pair of symbols, $\tau_a^{-1}\tau_b$ transforms a into b. This proves that G is transitive.

Since the order of a finite group is divisible by the index of any of its subgroups (Theorem 3, p. 35) we have the following useful corollary.

PROPOSITION 29. *The order of a transitive group of degree n is divisible by n.*

The concept of transitiveness can be generalized.

DEFINITION 13. *A group G of permutations is said to be **k-ply transitive** if it contains at least one permutation θ which changes any ordered set of k distinct objects $a_1, a_2, \ldots, a_k$ into any other such set $b_1, b_2, \ldots, b_k$ (the two sets may have elements in common); that is $a_i\theta = b_i$ $(i = 1, 2, \ldots, k)$.*

Clearly, if G is of degree n, then $k \leq n$. Also if G is k-ply transitive and if $l < k$, then G is *a fortiori* l-ply transitive.

The group S_n is k-ply transitive, where k is any one of the integers $1, 2, \ldots, n$.

Let v be the number of ordered sets comprising k objects chosen from all the n objects on which G acts. Then

$$v = n(n-1)\ldots(n-k+1).$$

Suppose now that G is k-ply transitive, and let H be the subgroup which leaves each of the objects

$$1, 2, \ldots, k$$

unchanged. By arguments similar to those used in the proof of Theorem 25 it can be shown that the index of H in G is equal to v, the cosets of H being in one-to-one correspondence with the v sets of k objects. Hence we have the result:

THEOREM 26. *The order of a k-ply transitive group of degree n is divisible by* $n(n-1)\ldots(n-k+1)$.

Alternatively, we might have developed the concept of multiple transitiveness inductively by using the following criterion.

PROPOSITION 30. *The group G is k-ply transitive if* (i) *G is simply transitive and* (ii) *the stabilizer* G_1 *is* $(k-1)$-*ply transitive with regard to the objects* $2, 3, \ldots, n$.

For example, in the case of A_4, which is exhibited in (7.28), the stabilizer of 1 is

$$G_1 : \iota, \ (234), \ (243).$$

Thus $[A_4 : G_1] = 12/3 = 4$, which confirms that A_4 is transitive. Now G_1 is transitive on 2, 3, 4; this may be verified directly, or alternatively it follows from the observation that the stabilizer of 2 in G_1, say G_{12}, reduces to ι and is therefore of index 3 in G_1. Since G_{12} fails to be transitive on the remaining objects 3, 4, we conclude that A_4 is (precisely) doubly transitive.

44. Primitive Groups. Let G be a transitive group and suppose that the n objects on which G acts can be arranged in an array of r rows and s columns, where

$$rs = n \quad (r > 1, s > 1),$$

thus

$$
\left.
\begin{array}{l}
a_1, \quad a_2, \ldots, a_s \\
b_1, \quad b_2, \ldots, b_s \\
 \cdot \qquad \cdot \ \cdots \ \cdot \\
k_1, \quad k_2, \ldots, k_s
\end{array}
\right\} \quad (r \text{ rows}) \qquad (7.46)
$$

in such a way that the permutations of G either rearrange the objects in any one row amongst themselves, or else interchange the objects of one row with those of another row (in some order). Thus two objects which stand in different rows of (7.46) are never transformed into objects of the same row, and conversely two objects of the same row are never sent into two different rows by the action of G. A transitive group which has this property is said to be **imprimitive** and the scheme (7.46) is called an **imprimitive system**. A group for which no imprimitive system exists, is said to be primitive. It should be noted that this concept applies only to transitive groups.

Example 1. The cyclic group $G = \text{gp } \{(1234)\}$, which consists of the permutations

$$\iota, \ (1234), \ (13)(24), \ (1432),$$

is imprimitive, having the imprimitive system

$$
\left.
\begin{array}{l}
13 \\
24
\end{array}
\right| .
$$

Indeed, the four permutations of G change this system into

$$
\left.\begin{array}{l}13\\24\end{array}\right|, \quad
\left.\begin{array}{l}24\\31\end{array}\right|, \quad
\left.\begin{array}{l}31\\42\end{array}\right|, \quad
\left.\begin{array}{l}42\\13\end{array}\right|
$$

respectively.

Example 2. A group may possess more than one imprimitive system. Thus in the case of the four-group

$$\iota, \ (12)(34), \ (13)(24), \ (14)(23),$$

each of the arrays

$$
\left.\begin{array}{l}12\\34\end{array}\right|, \quad
\left.\begin{array}{l}13\\24\end{array}\right|, \quad
\left.\begin{array}{l}14\\23\end{array}\right|
$$

can serve as an imprimitive system.

A doubly-transitive group is always primitive. For a doubly-transitive group would have to contain a permutation which sends the pair a_1, a_2 into the pair a_1, b_2. This would, however, be incompatible with the existence of an imprimitive system such as (7.46).

In particular all symmetric groups S_n are primitive.

45. Symmetry Groups. Let Σ be a finite or infinite set of points situated in a three-dimensional Euclidean space with origin 0. Any rotation about an axis through 0 which transforms Σ into itself, is called a symmetry of Σ with respect to 0. It follows from §6 of Chapter I (p. 18) that the symmetries of Σ form a group under composition of maps. If there are no non-trivial rotations which bring Σ into coincidence with itself, then the symmetry group reduces to the identity transformation.

In this section we shall discuss the symmetry groups for some geometrical configurations, including those of the five regular solids. The groups which emerge are already known to us.

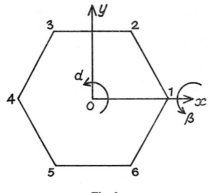

Fig. 3

(i) *Dihedral groups.* Consider a plane lamina having the shape of a regular polygon of n vertices, and suppose that the two sides of the lamina are completely alike (Fig. 3 illustrates the case in which $n = 6$). We choose the coordinate axes in such a way that the lamina lies in the (x, y)-plane, with its centre at the origin and the x-axis passing through a particular vertex, which we label 1. There are $2n$ rotations, including the identical operation, which bring the lamina into coincidence with itself. In the first place, if α denotes the rotation through $2\pi/n$ about the z-axis, we have the n symmetry operations

$$\iota\,(=\alpha^0), \quad \alpha, \quad \alpha^2, \ldots, \alpha^{n-1},$$

where

$$\alpha^n = \iota. \tag{7.47}$$

A further symmetry operation, β, consists in reversing the two sides of the lamina. This may be accomplished by a rotation through π about the x-axis. (The coordinate axes are supposed to be fixed in space). Clearly

$$\beta^2 = \iota, \tag{7.48}$$

because β^2 corresponds to a rotation through 2π and is therefore equal to the identity operation. Now the $2n$ operations

$$\alpha^k \beta^l \ (k = 0, 1, \ldots, n-1; \quad l = 0, 1)$$

constitute all the symmetries of the lamina; for they allow any vertex to be brought into the position of any other vertex, with or without the reversal of the two faces. In order to determine the structure of the symmetry group we have to find a relation between the operations α and β. A simple geometrical consideration shows that

$$\alpha\beta = \beta\alpha^{-1},$$

which, by virtue of (7.48), is equivalent to

$$(\alpha\beta)^2 = \iota \tag{7.49}$$

(The reader is recommended to verify this by drawing diagrams analogous to those on p. 8). Our result may be summarized as follows:

The symmetry group of a regular n-gonal lamina is the dihedral group of order 2n, given by the defining relations

$$\alpha^n = \beta^2 = (\alpha\beta)^2 = \iota. \tag{7.50}$$

We recall that this group was mentioned in Chapter II, exercise 7 (p. 55).

It is of interest to obtain analytical expressions for the operations of the dihedral group. Let x be a variable ranging over the integers $1, 2, \ldots, n$, which denote the vertices of the lamina in the counterclockwise order. The operation α is described by the congruence relation

$$x\alpha \equiv x+1 \pmod{n}. \tag{7.51}$$

Again, if we write $x = 1 + z$, then the image of x under β is $1 - z$. Thus

$$x\beta \equiv 2 - x \pmod{n}. \tag{7.52}$$

All relations between the generating elements α and β may be derived from (7.51) and (7.52); for example, we have that

$$x\alpha\beta \equiv (x+1)\beta \equiv 2 - (x+1) \equiv 1 - x,$$

$$x(\alpha\beta)^2 \equiv (1-x)\alpha\beta \equiv 1 - (1-x) \equiv x,$$

which confirms that $(\alpha\beta)^2 = \iota$.

(ii) *The tetrahedral group.* This is the name given to the symmetry group of a regular tetrahedron which is free to rotate about its centre 0. There are twelve rotations which bring the tetrahedron into coincidence with itself. First we select four operations which carry the vertex 1 into the position of any one of the vertices 1, 2, 3 or 4. Thereafter, if 1 occupies the position x, the solid can be rotated through an angle 0 or $2\pi/3$ or $4\pi/3$ about the line $0x$, whereby the three faces meeting at x are cyclically interchanged. Thus we have $4 \times 3 = 12$ operations in all.

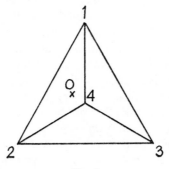

Fig. 4

The operations of the tetrahedral group permute the four vertices in some way; the group is therefore isomorphic with a subgroup of S_4. When one vertex is fixed, the remaining three vertices, say a, b, c, can be permuted cyclically. Hence the tetrahedral group contains all cycles $(a\ b\ c)$. By Proposition 27, these cycles generate the alternating group A_4. Since both groups are of order 12, we have proved that the *tetrahedral group is isomorphic with A_4.*

F

(iii) *The octahedral (hexahedral) group.* The centres of the faces of a regular octahedron may be regarded as the vertices of a cube (hexahedron), and conversely to every cube we can inscribe an octahedron whose vertices lie at the centres of the faces of the cube. Hence these two solids have the same symmetry, that is if one is transformed into itself, so is the other. Thus the octahedral and hexahedral groups are identical, though only the first name is in common use. In the present discussion we find it more convenient to consider the symmetry of the cube rather than that of the octahedron.

We observe that the group of the cube consists of 24 operations. For, in the first place, a given vertex may be brought into the position of any one of the eight vertices. When this has been done, the solid can be rotated through one of the angles 0, $2\pi/3$ or $4\pi/3$ about the diameter through this vertex, giving in all $8 \times 3 = 24$ rotations, including the identity.

The cube has four diameters (lines through 0 joining a pair of 'diametrically' opposite vertices). When the cube is transformed into itself, these four diameters are permuted in some manner. Thus the group of the cube is homomorphically mapped into S_4. Next, we determine the kernel of this homomorphism. If one particular diameter is carried over into itself, then either this diameter coincides with the axis of rotation, or else the two end-points of the diameter are interchanged; in the latter case, the axis of rotation is at right-angles to the diameter, and the angle of rotation is equal to π. A rotation belonging to the kernel would have to turn each of the four diameters into itself. The axis of this rotation would therefore have to be at right-angles to at least three of the diameters. This is evidently impossible, unless the operation is the identity. Hence the kernel is trivial, and *the octahedral group is isomorphic with S_4.*

(iv) *The icosahedral (dodecahedral) group.* Turning now to the last two of the regular polyhedra, we observe that the icosahedron and dodecahedron have the same symmetry. For the centres of the twenty faces of an icosahedron may be joined to form a regular dodecahedron; and, conversely, the centres of the six faces of a cube may be regarded as the vertices of an icosahedron. Thus the icosahedral and dodecahedral groups are identical. Either solid may be used to examine the structure of this group. We decide to choose the dodecahedron.

First, we remark that the dodecahedral group consists of 60 operations. For any vertex may be brought into the position of any one of the 20 vertices. When the vertex has reached its final position, the solid may be rotated about the diameter through it. This operation causes cyclic interchanges of the three faces that meet at the extremities of the diameter. Thus the possible angles of rotation are $0, 2\pi/3$ or $4\pi/3$. It follows that there are in all $20 \times 3 = 60$ operations, including the identity, which bring the dodecahedron into coincidence with itself.

Next we seek a faithful permutation representation of the dodecahedral group. It will turn out that the group is isomorphic with a subgroup of S_5. Thus we shall describe five objects which are permuted when the dodecahedron is rotated into itself. According to

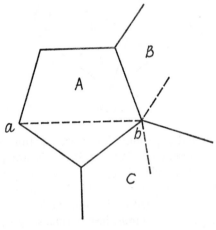

Fig. 5

Euclid's classical construction* a cube may be inscribed in the dodecahedron as follows: select a face A and draw a diagonal, say ab, in it (a diagonal is a line joining two non-adjacent vertices of a face). At the point b, the face A abuts on two further faces, say B and C. Then it can be shown that both in B and in C there is precisely one diagonal which is at right-angles to ab, and these new diagonals are at right-angles to each other. The construction is now repeated with the diagonals in B and C referred to; at the other extremity of either diagonal we determine two further diagonals in the adjacent faces to make a right-angled tripod, and so on. (The validity of these assertions is best verified by the inspection of a model). Thus,

*Elements, Book XIII, Proposition 17.

starting from *ab*, we have singled out a unique diagonal in each of the twelve faces, and these diagonals form the edges of a cube inscribed in the dodecahedron. Now each face has five diagonals (Fig. 6), and we could have begun the above construction with any one of these diagonals. Thus five cubes can be inscribed, and these objects are permuted in any symmetry operation upon the dodecahedron. We

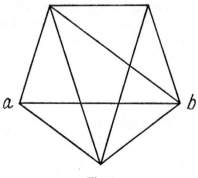

Fig. 6

have therefore found a permutation representation of degree five. Moreover, this representation is faithful; that is, any rotation which turns each of the five cubes into itself, necessarily reduces to the identity (we ask the reader to accept this fact without further proof). It follows that the dodecahedron is isomorphic with a subgroup of S_5; since it is of index two, it must be a normal subgroup (remark iv., p. 62), whence we infer from Proposition 28, that *the dodecahedral (icosahedral) group is isomorphic with A_5.*

Exercises

(1) Show that $(ab \ldots lx)(x\alpha\beta \ldots \lambda) = (ab \ldots l\alpha\beta \ldots \lambda x)$, where $a, b, \ldots, l, x, \alpha, \beta, \ldots, \lambda$ are distinct symbols.

(2) Prove that a permutation of degree n which is the product of r mutually exclusive cycles (including those of order 1) is even or odd according as $n - r$ is even or odd.

(3) Show that S_n may be generated by the transpositions
$(12), (23), \ldots, (n-1, n)$.

(4) Show that S_n may be generated by the permutations
$\gamma = (1\ 2\ \ldots\ n)$ and $\tau = (12)$.

(5) Prove that a regular permutation can be expressed as the power of a cycle and that, conversely, if $\gamma = (1\ 2\ \ldots\ m)$, then γ^s is a regular permutation consisting of d cycles of degree r, where $d = (m, s)$ and $r = m/d$.

(6) Prove that the centralizer of $\gamma = (a_1, a_2, \ldots, a_n)$ in S_n consists of $\iota, \gamma, \gamma^2, \ldots, \gamma^{n-1}$.

(7) Prove that when $n > 2$, the centralizer of $\lambda = (a_1, a_2, \ldots, a_{n-1})$ in S_n consists of $\iota, \lambda, \lambda^2, \ldots, \lambda^{n-2}$.

(8) Prove that, when $n > 2$, the centre of S_n consists of the identity permutation alone.

(9) The **left regular representation** of a group G is defined as follows: corresponding to a fixed element u of G there is a permutation λ_u, acting on the elements of G in accordance with the rule $a\lambda_u = u^{-1}a$ ($a \in G$). Verify that (i) $\lambda_u\lambda_v = \lambda_{uv}$; (ii) $\lambda_u = \iota$, if and only if $u = 1$; (iii) $\lambda_u\rho_x = \rho_x\lambda_u$, where ρ_x is defined in (7.42); (iv) if θ is a permutation of the elements of G which commutes with all the λ_u, then $\theta = \rho_x$ for some x, and if η commutes with all the ρ_x then $\eta = \lambda_u$ for some u.

(10) Prove that if G is a simple group of order 168 and H is a proper subgroup of G, then $[G:H] \geq 6$.

(11) Obtain the symmetry group of a rectangular (non-square) lamina.

(12) Prove that when the g elements of a transitive group of degree n are written as products of mutually exclusive cycles of degrees greater than one, then they involve between them $(n-1)g$ letters.

F*

VIII. Sylow's Theorems

46. Prime-power Subgroups. Lagrange's theorem states that if G is a finite group of order g, then the order of a subgroup of G must divide g. The converse of this theorem is false; for we have seen that there are groups (example, p. 146) which do not possess subgroups corresponding to certain divisors of g. However, if p^b is a power of a prime p such that p^b divides g, then G has at least one subgroup of order p^b. This remarkable fact was discovered in 1872 by the Norwegian mathematician L. Sylow. It is of far-reaching consequence in the theory of groups, and it provides one of the most striking examples of the subtle connections between arithmetical and structural properties of a group. Several proofs of L. Sylow's celebrated results can be found in the literature. We present* here an elegant line of argument which is due to H. Wielandt (1959); it proceeds from first principles and uses only some elementary ideas on permutations.

THEOREM 27. *Let G be a finite group of order g and suppose that p is a prime such that p^b divides g, where b is a positive integer. Then G possesses m subgroups of order p^b, where m is a positive integer satisfying $m = 1 \pmod{p}$.*

Proof. (1) Write

$$g = p^b z, \tag{8.1}$$

where z is a positive integer, which need not be coprime with p. Make a complete list, $\mathscr{K}$, of all subsets consisting of p^b distinct elements of G. Thus if there are n such subsets, we write

$$\mathscr{K} : K_1, K_2, \ldots, K_n \tag{8.2}$$

In fact, n is equal to the binomial coefficient $\binom{g}{p^b}$; but this information will not be required subsequently. The theorem asserts that at least one of the subsets (8.2) is a subgroup.

---

* Our exposition follows B. Huppert, *Endliche Gruppen* I, (Springer 1967), p. 33.

A subset K belongs to $\mathscr{K}$ if and only if, in the notation of p. 30,

$$|K| = p^b.$$

If x is any element of G, then $|Kx| = |K|$. Hence Kx also belongs to $\mathscr{K}$. Indeed, the map

$$K_i \rightarrow K_i x \quad (i = 1, 2, \ldots, n)$$

constitutes a permutation of $\mathscr{K}$. We say that, in this sense, G acts on $\mathscr{K}$. With regard to this action an equivalence relation on $\mathscr{K}$ may be defined as follows: the subsets K_i and K_j are said to be *equivalent*, if there exists an element x of G such that $K_i = K_j x$. The reader will have no difficulty in verifying that the usual axioms of an equivalence relation are satisfied. As a consequence, $\mathscr{K}$ is divided into mutually disjoint equivalence classes or *orbits*, as they are called in this context. Thus the orbit of K, which we shall denote by $o(K)$, consists of all subsets of the form Kx, $(x \in G)$. As x runs through G, each element of the orbit will, in general, be obtained several times. The numbers of distinct subsets contained in $o(K)$ will be written $|o(K)|$. The decomposition of $\mathscr{K}$ into orbits is then expressed as

$$\mathscr{K} = o(K) \cup o(K') \cup o(K'') \cup \ldots, \qquad (8.3)$$

where $K, K', K'', \ldots$ is a set of representatives of the orbits. Counting the number of elements on both sides we obtain that

$$n = |o(K)| + |o(K')| + |o(K'')| + \ldots. \qquad (8.4)$$

(2) We shall now investigate one of the orbits, say $o(K)$, in more detail. Let S be the stabilizer of K under the action of G, that is

$$S = \{u \in G \mid Ku = K\}.$$

The reader will easily verify that S is a subgroup of G (see p. 147). Suppose that

$$G = \bigcup_{i=1}^{r} St_i \quad (t_1 = 1)$$

is the right coset decomposition of G relative to S. We claim that $o(K)$ consists of the subsets

$$Kt_1, Kt_2, \ldots, Kt_r. \qquad (8.5)$$

Obviously all these sets belong to $o(K)$, and they are distinct; for if $Kt_i = Kt_j$, it would follow that $Kt_i t_j^{-1} = K$, that is $t_i t_j^{-1} \in S$ and therefore $St_i = St_j$, which implies that $i = j$. Next, an arbitrary member of $o(K)$ is of the form Kx. If x lies in the coset St_i, we have that $x = ut_i$, where $u \in S$, and hence $Kx = Kut_i = Kt_i$. Thus we have proved that

$$|o(K)| = [G:S]. \tag{8.6}$$

Further information about S can be gleaned from the fact that K is of prime-power cardinal. The defining property of the stabilizer can be expressed by the equation

$$KS = K,$$

viewed as a relation between subsets of G. More precisely, if $K = v_1 \cup v_2 \cup v_3 \ldots$, we have that

$$K = v_1 S \cup v_2 S \cup v_3 S \cup \ldots. \tag{8.7}$$

Thus K is the union of left cosets of S. We know that two such cosets are either disjoint or identical and that each contains $|S|$ elements. Thus if the number of distinct cosets in (8.7) is equal to f, we have that

$$p^b = f|S|.$$

It follows that $|S|$ is a power of p, say

$$|S| = p^c, \tag{8.8}$$

where $c \leqq b$. Now two cases have to be distinguished.

(i) $|S| = p^b$. We do not know yet whether this case can arise. But if it does, then

$$|o(K)| = g/p^b = z,$$

where z is defined in (8.1). As $|S|$ now takes its largest value, we may term $o(K)$ a *minimal orbit*. Since by the present hypothesis K and S have the same cardinal, we infer from (8.7) that K reduces to a single coset, say

$$K = vS \quad (v \in K).$$

The subset

$$H = Kv^{-1} = vSv^{-1}$$

clearly belongs to $o(K)$ and, moreover, is a subgroup, namely a group conjugate with S. Thus we have reached the conclusion that every minimal orbit contains at least one subgroup.

Since $|H| = p^b$, it follows that

$$[G:H] = z = |o(K)|.$$

Let

$$Hw_1, Hw_2, \ldots, Hw_z \tag{8.9}$$

be the cosets of H in G. Each of these z cosets belongs to $o(K)$, because H does; and since they are distinct, they constitute the whole of $o(K)$. But we know that precisely one of the cosets, namely H, is a group. Hence we have shown that *a minimal orbit contains one and only one subgroup of G.*

(ii) $|S| = p^c < p^b$. In this case the orbit $o(K)$ is not minimal and

$$|o(K)| = g/p^c = zp^{b-c}$$

Hence

$$|o(K)| \equiv 0 \; (\mathrm{mod}\, pz). \tag{8.10}$$

A non-minimal orbit cannot contain a subgroup; for if it did, we could choose this subgroup as the generator of $o(K)$ and hence without loss of generality assume that K itself was a group. Then K would lie in its own stabilizer, because $KK = K$ (see (2.6), p. 31). Thus $|S| \geq |K| = p^b$, which is incompatible with the hypothesis (ii).

(3) Returning to (8.4) we shall separate the minimal terms, if any, from the others. There is precisely one subgroup in each minimal orbit; and distinct orbits contain distinct subgroups, because the orbits are disjoint. The cardinal $|o(K)|$ is equal to z for each minimal orbit and the number of such orbits is equal to m, the integer defined in the theorem. (Note, however, that at this stage we still do not know whether m is positive.) Hence the total contributions to (8.4) from all the minimal orbits is equal to mz. Since, by (8.10), each of the remaining terms in (8.4) is divisible by pz, we can summarize the situation by the congruence

$$n \equiv mz \; (\mathrm{mod}\, pz). \tag{8.11}$$

It is a crucial feature of this proof that the number n, which we defined on p. 158, depends only on the order of the group G but not on its structure. Thus n has the same value for all groups of order $p^b z$, whilst m varies for fixed n. We should therefore write (8.11) more explicitly as

$$n = m_G z + k_G p z,$$

where m_G and k_G are integers depending on G. In order to obtain information about n we apply this result to the cyclic group C of order

$p^b z$. We know from Theorem 4 (p. 37) that C has precisely one subgroup of order p^b. Thus $m_C = 1$ and hence

$$n = z + k_C p z.$$

Equating the two expressions for n we find that

$$z + k_C p z = m_G z + k_G p z,$$

whence on dividing throughout by z,

$$m_G \equiv 1 \pmod{p},$$

as claimed.

47. Sylow's Theorems. Sylow's results are usually presented in three theorems, which we shall give in this section.

THEOREM 28. **(Sylow's First Theorem):** *If p^a is the highest power of a prime p dividing the order of a group G, then G possesses at least one subgroup of order p^a.*

Proof. This is a special case of Theorem 27. It corresponds to the greatest possible value of the exponent b.

DEFINITION 14. *Let G be a finite group of order g. Suppose that $g = p^a g'$, where p is a prime and $(g', p) = 1$. Then any subgroup of G of order p^a is called a* **Sylow p-group** *of G.*

A group G may possess more than one Sylow group corresponding to the same prime. Indeed, if P is a subgroup of order p^a, so is $x^{-1}Px$ where x is an arbitrary element of G. In other words, the conjugate of a Sylow group is also a Sylow group. Of course, conjugate groups need not be distinct. But the next theorem tells us that no other Sylow groups can exist.

THEOREM 29. **(Sylow's Second Theorem).** *All Sylow groups of G belonging to the same prime are conjugate with one another in G.*

Proof. As in Definition 14, put $|G| = g = p^a g'$, where $(g', p) = 1$. Suppose that A and B are subgroups of order p^a. We use the double coset decomposition of G relative to A and B (Theorem 6, p. 54), thus in the present case

$$G = At_1 B \cup At_2 B \cup \ldots \cup At_r B,$$

$$g = p^{2a} \sum_{i=1}^{r} d_i^{-1} \tag{8.12}$$

$$d_i = |t_i^{-1} At_i \cap B|. \tag{8.13}$$

Dividing (8.12) throughout by p^a we obtain that

$$g' = p^a \sum_{i=1}^{r} d_i^{-1}. \tag{8.14}$$

Now d_i is the order of a subgroup of B and hence must be equal to a non-negative power of p. Hence each term on the right of (8.14) is either equal to unity or else is a power of p with positive exponent. But g' is not divisible by p. Therefore at least one of the terms on the right must be equal to one, say $p^a d_j^{-1} = 1$, that is $d_j = p^a$. We then have that

$$p^a = |t_j^{-1} A t_j \cap B|.$$

Since the groups $t_j^{-1} A t_j$ and B are both of order p^a, their intersection can be of order p^a only if the groups are identical. Thus

$$B = t_j^{-1} A t_j,$$

that is A and B are conjugate, as was asserted.

COROLLARY 1. *A finite group G possesses a unique Sylow group P corresponding to a given prime p if and only if P is normal in G.*

Proof. The condition of uniqueness is equivalent to the statement that $x^{-1} P x = P$ for all x in G; but this means that P is a normal subgroup.

In the case of finite Abelian groups the Sylow groups are necessarily unique. The concept of a Sylow group coincides with that of a p-primary component (p. 96). In the multiplicative nomenclature, Theorem 16 (p. 96) may be reformulated as follows.

COROLLARY 2. *A finite Abelian group is the direct product of its Sylow groups.*

The next theorem gives more precise information about the number of Sylow p-groups.

THEOREM 30: **(Sylow's Third Theorem).** *Let r be the number of Sylow p-groups of G. Then r is an integer of the form $1 + pk$ and is a factor of the order of G.*

Proof. The fact that $r \equiv 1 \pmod{p}$ has already been established in Theorem 27. It remains to show that $r | g$, where $g = |G|$. Let

$$\mathscr{P} : P_1 (=P), P_2, \ldots, P_r$$

be the set of all Sylow p-groups of G. Then, by Theorem 29, $\mathscr{P}$ is a complete set of conjugates of P. The reader who has mastered exercise 6 of Chapter III (p. 79) will know that

$$r = [G:N(P)], \tag{8.15}$$

where $N(P)$ is the normalizer if P in G. Thus, if $|N(P)| = n$, then $g = nr$, which shows that $r \mid g$. The relation (8.15) is analogous to (8.6). Indeed, we may define an action of G on the set $\mathscr{P}$ by associating with an arbitrary element x of G the map

$$P \to x^{-1}Px \quad (P \in \mathscr{P}),$$

which causes a permutation of $\mathscr{P}$. As x runs through G, any member of $\mathscr{P}$ will be obtained, that is the whole of $\mathscr{P}$ is the orbit of P, and we have that

$$|o(P)| = r.$$

The stabilizer of P consists of those elements u of G for which $u^{-1}Pu = P$. Thus in our present context the stabilizer becomes the normalizer. On writing $N(P)$ for S, we see that (8.6) reduces to (8.15).

48. Applications and Examples. Sylow's theorems provide powerful tools for examining the structure of a finite group. The method is particularly effective when the group possesses a unique Sylow group for some prime.

PROPOSITION 31. *Let G be of order pq, where p and q are primes such that $p < q$ and $q \not\equiv 1 \pmod{p}$. Then G is necessarily Abelian.*

Proof. Let r be the number of Sylow p-groups. By Theorem 30, $r \mid pq$ and $r = 1+pk$. Evidently, $(r, p) = 1$ and hence $r \mid q$. Since q is a prime, it follows that either $r = 1$ or $r = q$. The latter alternative would mean that $q = 1+pk$, that is $q \equiv 1 \pmod{p}$, which we have excluded by hypothesis. Hence by Corollary 1, G possesses a normal subgroup P of order p, which is necessarily cyclic. We denote its generator by u. Thus

$$P \lhd G, \quad P = \mathrm{gp}\{u\}. \tag{8.16}$$

Next suppose that G has s Sylow q-groups. Then $s \mid pq$ and $s = 1+ql$. Since $(s, q) = 1$, we must have that $s \mid p$, and hence $s \leq p$. If $l \geq 1$, then $s \geq 1+q > p$, a contradiction. It follows that $l = 0$, and G has a normal subgroup Q of order q with generator v, say. Thus

$$Q \lhd G, \quad Q = \mathrm{gp}\{v\}. \tag{8.17}$$

Since P and Q are of coprime orders,

$$P \cap Q = \{1\}. \tag{8.18}$$

It follows from Proposition 11 (p. 75) that the elements of P and Q commute in pairs. In particular,

$$uv = vu. \tag{8.19}$$

The products

$$u^\alpha v^\beta \ (\alpha = 0, 1, \ldots, p-1; \quad \beta = 0, 1, \ldots, q-1)$$

are distinct, since an equality between them would contradict (8.18). Hence these elements constitute the whole group, and (8.19) makes it plain that the group is Abelian.

Example 1. *There can be no simple group of order* 200.

For since $200 = 5^2 \times 8$, the group contains r Sylow groups of order 25, where r is of the form $1 + 5k$ and a divisor of 200. Since $(r, 5) = 1$, we must have $r \mid 8$, which is impossible unless $k = 0$. Hence the group contains a unique normal subgroup of order 25 and is therefore not simple.

Example 2. *There can be no simple group of order* 30.

For if there were such a group, none of the Sylow groups would be unique. Hence there would be $1 + 5 \ (= 6)$ distinct Sylow groups of order 5 comprising $6 \times 4 \ (= 24)$ elements of order 5. Similarly, there would be $1 + 3 \times 3 \ (= 10)$ distinct Sylow groups of order 3, which would yield 20 elements of order 3. Thus the total number of elements would exceed 30.

We continue with a more general result about Sylow groups.

THEOREM 31. *Let P be a Sylow subgroup of a finite group G and suppose that H is a subgroup of G which contains the normalizer of P. Then H is its own normalizer.*

Proof. Let $u \in N(H)$, the normalizer of H, that is $u^{-1}Hu = H$. Now $P \leq N(P) \leq H$, and hence $u^{-1}Pu \leq u^{-1}Hu = H$. Thus $u^{-1}Pu$, being of the same order as P, is also a Sylow group of H. Applying Theorem 29 to H we infer that there exists an element h_1 of H such that

$$h_1^{-1}(u^{-1}Pu)h_1 = P.$$

This means that uh_1 normalizes P. Since, by hypothesis, $N(P) \leq H$, it follows that $uh_1 = h_2$, where $h_2 \in H$. Hence $u \in H$, which proves the theorem. Finally, we shall show that the property mentioned in the Corollary |2| p. 163 is in fact characteristic for all finite nilpotent groups.

THEOREM 32. *Let G be a finite group of order $p_1^{\alpha_1} p_2^{\alpha_2} \ldots p_r^{\alpha_r}$ and let $P_1, P_2, \ldots, P_r$ be a set of Sylow groups of G corresponding to the primes $p_1, p_2, \ldots, p_r$ respectively. Then G is nilpotent if and only if*

$\qquad$ (i) $P_i \lhd G$ $(i = 1, 2, \ldots, r)$, and

$\qquad$ (ii) $G = P_1 \times P_2 \times \ldots \times P_r$. $\qquad\qquad$ (8.20)

Proof. First, assume that (8.20) holds. We know that each factor in a direct product is a normal subgroup (p. 75). Also, by example 2, p. 126, each Sylow group is nilpotent. It remains to show that a direct product of nilpotent groups is itself nilpotent. Thus suppose that K and L are nilpotent and consider the group $K \times L$. If $\Gamma_i(K)$, $\Gamma_i(L)$, $\Gamma_i(k \times L)$ are typical terms of the series (6.18) for the groups K, L and $K \times L$ respectively, then it is clear that

$$\Gamma_i(K \times L) = \Gamma_i(K) \times \Gamma_i(L) \quad (i = 1, 2, \ldots).$$

Hence if $\Gamma_i(K)$ and $\Gamma_i(L)$ reduce to the unit group for sufficiently great values of i, then so does $\Gamma_i(K \times L)$, that is $K \times L$ is nilpotent. Therefore (8.20) implies the nilpotency of G.

Conversely, suppose that G is a finite nilpotent group. Let P be a Sylow group of G corresponding to a particular prime and put $H = N(P)$. We assert that $H = G$, that is $P \lhd G$. For if, on the contrary, H were a proper subgroup then by Proposition 20 (p. 27), $N(H) > H$; on the other hand, by Theorem 31, $N(H) = H$. This contradiction shows that $H = G$. Hence $P_i \lhd G$ $(i = 1, 2, \ldots, r)$.

Obviously $P_i \cap P_j = \{1\}$ when $i \neq j$. Hence by Proposition 11 (p. 75) and by the definition of the interior direct product (p. 43)

$$P_1 P_2 \ldots P_r = P_1 \times P_2 \times \ldots \times P_r.$$

This is a subgroup of order $p_1^{\alpha_1} p_2^{\alpha_2} \ldots p_r^{\alpha_r}$ and therefore coincides with G.

Exercises

(1)   Show that A_4 has one Sylow group of order 4 and four Sylow groups of order 3.

(2)   Obtain one of the Sylow 2-groups of S_4. With which of the groups given on pp. 45–46 is it isomorphic? How many Sylow 2-groups are there?

(3) Prove that there is no simple group of order 56.

(4) Let G be a group of order p^2q, where p and q are primes such that q is less than p and is not a factor of $p^2 - 1$. Prove that G is Abelian.

(5) Let p be a prime which divides the order of a group G. Prove that if K is a subgroup of G such that $|K|$ is a power of p, then K is contained in at least one Sylow p-group.

(6) Show that a normal p-subgroup is contained in every Sylow p-subgroup.

(7) Let P be a Sylow p-subgroup of a finite group G and suppose that H is a normal subgroup of G. Prove that (i) HP/H is a Sylow p-subgroup of G/H and (ii) $H \cap P$ is a Sylow p-subgroup of H.

Answers to Exercises

Chapter I

(2) the associative law fails.

(3) $xa^n = \alpha^n x + \beta(\alpha^n - 1)/(\alpha - 1)$.

(4) $ab = (ab)^{-1} = b^{-1}a^{-1} = ba$.

(5) $ba = a^{-1}(ab)a^{-1}$, see p. 17, (iii)

(6) note that $a^m b^{n-1} = b(ab^{-1})b^{-1}$, $a^{m-2}b^n = a^{-1}(a^{-1}b)a$.

(8) there exist integers u, v such that $um + vn = 1$; put $y = x^{vn}$, $z = x^{um}$.

(9) those elements which do not satisfy the equation $x^2 = 1$ can be grouped into disjoint pairs (u, u^{-1}), (v, v^{-1}), ... The equation therefore has an even number of solutions, one of which is 1.

(10) the orders are 1, 3, 6, 3, 6, 2 respectively and 3 or 5 can be taken as a generator.

(13) (i) (1478) (265) (39); (ii) $(acdf)$ (be).

(14) (i) $(abc \ldots k)$; (ii) $(a_r yb_1 \ldots b_s xc_1 \ldots c_t)$; (iii) $(a_r yc_1 c_2 \ldots c_t)$ $(xzb_1 b_2 \ldots b_s)$.

Chapter II

(2) If $u, v \in At \cap Bs$, then $uv^{-1} \in A \cap B$, and therefore $Du = Dv$. The number of distinct cosets of D cannot exceed the number of non-empty intersections $At \cap Bs$, and every element of G lies in one of these intersections.

(3) Since $|A \cap B|$ divides both $|A|$ and $|B|$, it follows that $|A \cap B| = 1$.

(4) If G is cyclic, the result follows from Theorem 4. If G is not cyclic, let $x \in G$ and $x \neq 1$; then gp $\{x\}$ is a proper subgroup.

(5) gp $\{a\}$, gp $\{a^2, b\}$, gp $\{ab, a^3b\}$

(6) It is sufficient to show that these relations imply those given on p. 47, thus $a = cd$, $ac = cdc$, whence $a^3 = 1$, $(ac)^2 = 1$.

(9) Write the elements in the form a^k, $a^k b(0 \leq k \leq 5)$, $c = b^{-1}ab$ must be a power of a and is of the same order as a. Since $c = a$ is excluded, it follows that $c = a^{-1}$. Again, $b^2 = a^l$ for suitable l and hence $b^2 = b^{-1}b^2b = b^{-1}a^lb = a^l$. Therefore $a^{2l} = 1$, whence $l = 0$ or $l = 3$.

(10) e.g., gp $\{2\} \times g\{-1\}$, that is gp $\{2\} \times$ gp $\{20\}$.

Chapter III

(1) If $b = t^{-1}at$, then $a^m = 1$ implies that $b^m = 1$, and conversely.

(2) (i) If $C(a)$ is the centralizer of a, then $C(a) = C(a^{-1})$, whence the result follows from Proposition 7. (ii) Use the class equation (3.5), where it may be assumed that $h_1 = 1$; if the assertion were false, the

168

remaining terms could be grouped into pairs of equal terms, each pair corresponding to inverse classes. But then g would be odd, contrary to hypotheses.

(3) Let $a = (a_{ij}) \in Z$, the centre of G. Then $ax = xa$ for all $x \in G$. In particular we may take $x = \text{diag}(x_1, x_2, \ldots, x_n)$, a diagonal matrix with distinct diagonal entries. It follows that $a_{ij}x_j = x_i a_{ij}$, whence $a_{ij} = 0$ when $i \neq j$; thus a is itself a diagonal matrix. Next, take for x the permutation matrix $p = (p_{ij})$, where $p_{i,i+1} = 1 (i < n)$, $p_{n,1} = 1$ and all other $p_{ij} = 0$. The equation $ap = pa$ implies that $a_{11} = a_{22} = \ldots = a_{nn}$, that is a is a scalar matrix.

(4) $Z = \text{gp}\{a^2\}$. The elements of G/Z are Z, Za, Zb, Zab, $G/Z \cong V$ (p. 46).

(5) If s and t are upper-triangle matrices, then st is an upper-triangle matrix with diagonal $s_{11}t_{11}, s_{22}, t_{22}, \ldots, s_{nn}t_{nn}$, whence the group properties of T are easily established. Let $\theta : T \rightarrow D$ be the map defined by $t\theta = \text{diag}(t_{11}, t_{22}, \ldots, t_{nn})$. Then E is the kernel of θ, and the assertion follows from the First Isomorphism Theorem.

(6) Note that $x^{-1}Hx = y^{-1}Hy$ if and only if $xy^{-1} \in N(H)$, that is $N(H)x = N(H)y$. (See the proof of Proposition 7, p. 58.)

(7) G/N is a finite group of order n, and the element Nt of G/N is of order h. Hence, by Lagranges Theorem, $h|n$. Also $(Nt)^r = Nt^r = N$, whence $h|r$ (Proposition 1, p. 17).

(8) Both parts are proved by induction on k, using that $ab = bac$, $ac = ca$ and $bc = cb$.

(9) By the Third Isomorphism Theorem $A/A \cap N \cong NA/N$, which is a subgroup of the finite group G/N of order n. Hence $|A/A \cap N|$ divides n.

(10) In each of these groups the centre is $Z = \text{gp}\{a^2\}$. As $a^2 = [a, b]$, $a^2 \in G'$ and hence $Z \leq G'$. On the other hand, G/Z is of order 4 and hence Abelian. Therefore (Theorem 11), $G' \leq Z$, whence $G' = Z = \text{gp}\{a^2\}$.

(11) Let $N \triangleleft G$ and $C = C(N)$ (p. 59). Thus if $c \in C$ then $cu = uc$ for all $u \in N$. If $t \in G$, then $c^t u^t = u^t c^t$; but u^t can equal any element of N. Hence $c^t \in C$, and so $C \triangleleft G$.

(12) $(xy)\theta = (xy)^{-1} = y^{-1}x^{-1} = x^{-1}y^{-1} = (x\theta)(y\theta)$. It is easy to show that θ is bijective.

(13) Let $x\tau = t^{-1}xt$ be an inner automorphism and α any automorphism; put $s = t\alpha$ and $x\sigma = s^{-1}xs$. Then $x\alpha\sigma = s^{-1}(x\alpha)s$, $x\tau\alpha = (t^{-1}xt) \times \alpha = s^{-1}(x\alpha)s$. Thus $\alpha\sigma = \tau\alpha$, $\alpha^{-1}\tau\alpha \in I(G)$, $I(G) \triangleleft A(G)$.

(14) Let $\alpha \in A(G)$; then $[a, b]\alpha = [a\alpha, b\alpha] \in G'$. Thus $G'\alpha \subset G'$. Similarly, $G'\alpha^{-1} \subset G'$. Thus $G' \subset G'\alpha$, whence $G'\alpha = G'$.

Chapter IV

(1) Construct a free Abelian group $F = \langle u_1, u_2, \ldots, u_n \rangle$ and put $v_1 = b_1 u_1 + b_2 u_2 + \ldots + b_n u_n$. There exist elements $v_2, v_3, \ldots, v_n$ such that $F = \langle v_1, v_2, \ldots, v_n \rangle$ and $v_i = \Sigma_j b_{ij}u_j$, where (b_{ij}) is a unimodular matrix with the desired property.

(2) Let $|A| = p_1 p_2 \ldots p_n$. In (4.47) the primary component P_i is order p_i and hence cyclic, say $P_i = \text{gp } \{x_i\}$. Then $x = x_1 x_2 \ldots x_n$ is an element of order $p_2 p_2 \ldots p_n$ and hence generates A.

(3) Let e_1 be the greatest invariant. The t_ρ are absent from (4.37), w_1 is of order e_1 and every element satisfies $e_1 x = 0$.

(4) Each of the ϕ (24) ($= 8$) residue classes satisfies $x^2 \equiv 1 \pmod{24}$.

(5) (i) 4, 3, 5; (ii) (4, 2), 3, (5, 5); 60, 10.

(6) $C_\infty \oplus C_3 \oplus C_6$.

(7) (i) $r = 1$, $e_1 = 2$; (ii) $r = 2$, $e_1 = 2$.

(8) $v_1 = u_1 + u_2 + k u_3$, $v_2 = u_2 - u_3$, $v_3 = -u_3$;
$s_1 = r_3$, $s_2 = r_2 - r_3$, $s_3 = r_1 + r_2 - (k+1) r_3$.
$e_1 = 1$, $e_2 = k - 1$, $e_3 = (k-1)(k+2)$.

(9) By virtue of Theorem 16 it suffices to prove the proposition for an Abelian prime-power group P. Let $|P| = p^m$. We have to show that, if $n \leq m$, there exists a subgroup P' such that $|P'| = p^n$. Suppose that $P = \Sigma_i \oplus P_i$, where $|P_i| = p^{\delta_i}$. Then $m = \Sigma_i \delta_i$. Evidently, we can write $n = \Sigma_i \lambda_i$ where $0 \leq \lambda_i \leq \delta_i$. By Theorem 4, there exists a subgroup $P_i' \subset P_i$ of order p^{λ_i}. Put $P' = \Sigma_i P_i'$.

(11) Let the elements of the group be identified with the p^3 vectors $a = (\alpha_1, \alpha_2, \alpha_3)$, where $0 \leq \alpha_i < p (i = 1, 2, 3)$. The first basis vector a_1 may be any one of the $p^3 - 1$ non-zero vectors. The second basis vector, a_2, can be any vector which is not a scalar multiple of a_1; there are $p^3 - p$ such vectors. Finally, a_3 completes the basis, provided that a_3 is not a linear combination of a_1 and a_2; there are $p^3 - p^2$ such vectors. Thus we have $(p^3 - 1)(p^3 - p)(p^3 - p^2)$ choices.

(12) Consider a free Abelian group $F = \langle u_1, u_2, \ldots, u_n \rangle$ and the subgroup $R = \text{gp } \{r_1, r_2, \ldots, r_m\}$, where $r_i = \Sigma_j b_{ij} u_j$. Changing the generators in F and R amounts to multiplying B on the right by Q and on the left by P.

Chapter V

(1) Let F be a free group, and let U be the collection of words with zero exponent sum for each generator. Then U is a subgroup and clearly $F' \subset U$. Conversely under the natural map $F \to F/F'$ each element of U is mapped into F'. Hence $U \subset F'$, so that $U = F'$.

(2) (i) $C_2 \times C_2$, (ii) C_4.

(3) The relations can be written $b^{-1} a^{-1} b a = b$, $a^{-1} b^{-1} a b = a$, whence on multiplying $ab = 1$. Therefore $b = a^{-1}$, and we find that $a = b = 1$.

Chapter VI

(1) In both cases $G \rhd \text{gp } \{a\} \rhd \text{gp } \{a^2\} \rhd \{1\}$ can be takena s a composition series. All composition factors are of order 2.

(2) Suppose that $G^{(s)} = \{1\}$. If $H \leq G$, then $H^{(i)} \leq G^{(i)}$ ($i = 1, 2, \ldots$). If $N \lhd G$, then $G/N = Gv$, where $v : G \to G/N$ is the natural epimorphism of G onto G/N. Note that $(Gv)^{(i)} = G^{(i)} v$ ($i = 1, 2, \ldots$). Hence the derived series for H and Gv terminate in the unit group after at most s steps.

(4) Since $\Gamma_3 = [\Gamma_2, G] = [G', G] = \{1\}$, G' lies in the centre. In formulae of ex (3), $[x, z]^y = [x, z]$, $(x, y]^z = [x, y]$.

(5) As in ex. (2).

(6) Since $\Gamma_4 = \{1\}$, $\Gamma_3 = [G', G]$ lies in the centre; in particular, $[v, x^{-1}] = c \in Z$, that is $v^{-1}xv = cx$. Similarly, if $y \in G$, $v^{-1}yv = dy$, where $d \in Z$. Now
$$v^{-1}[x, y]v = v^{-1}(x^{-1}y^{-1}xy)v = c^{-1}d^{-1}cd[x, y] = [x, y]$$
Hence is v commutes with each element of G'.

(7) By Proposition 20, $M < N(M)$; hence $N(M) = G$, that is $M \lhd G$. Also G/M cannot have a proper subgroup, as this group would contain M. Therefore $|G/M|$ is a prime.

(8) The elements of $D(2^n)$ can be expressed as $a^\alpha b^\beta (\alpha = 0, 1, \ldots, 2^n-1, \beta = 0, 1)$. Since $b^{-1}ab = a^{-1}$, a central element must satisfy $a^\alpha b^\beta = a^{-\alpha}b^\beta$, whence $\alpha = 0$ or 2^{n-1}, and $\beta = 0$ because b does not lie in the centre. Hence $Z_1 = \{1, a^{2^{n-1}}\}$. If $\bar{a} = aZ_1$, $\bar{b} = bZ_1$, then $(\bar{a})^{2^{n-1}} = \bar{b}^2 = (\overline{ab})^2 = \bar{1}$. Consecutive terms of the upper central series have index 2.

Chapter VII

(2) Let $\xi = \sigma_1\sigma_2 \ldots \sigma_r$, where σ_i is a cycle of degree $m_i(i = 1, 2, \ldots, r)$. Then $n = m_1 + m_2 + \ldots + m_r$. Using (7.25) we find that $\zeta(\xi) = (-1)^v$, where $v = \Sigma_i(m_i - 1) = n - r$.

(3) Observe that $(12)(23)(12) = (13)$, $(13)(34)(13) = (14)$, and so on. Then refer to Proposition 25.

(4) Consider $\gamma^{-r}\tau\gamma^r(r = 0, 1, \ldots, r-2)$ and use the previous exercise.

(5) The first part follows from the formula
$$\prod_{\lambda=1}^{k} (a_1^{(\lambda)}a_2^{(\lambda)} \ldots a_r^{(\lambda)})$$
$$= (a_1^{(1)}a_1^{(2)} \ldots a_1^{(k)}a_2^{(1)}a_2^{(2)} \ldots a_2^{(k)}a_3^{(1)} \ldots)^k$$

The second part is a consequence of Cayley's theorem and the fact that γ^s is of order m/d (see Proposition 2, p. 17).

(6) By Proposition 22 the conjugacy class of γ contains $(n-1)!$ elements. Hence $|C(\gamma)| = n$ (Proposition 7, p. 58). But $C(\gamma)$ certainly contains the n powers of γ and therefore no other elements.

(7) The conjugacy class of λ contains $n!/(n-1)$ elements. Hence $|C(\lambda)| = n-1$; but $C(\lambda)$ contains the $n-1$ powers of λ.

(8) If Z is the centre of S_n, then $Z < C(\gamma) \cap C(\lambda) = \{\iota\}$, where γ and λ are defined in exercises (6) and (7).

(9) (i) $a\lambda_u\lambda_v = u^{-1}a\lambda_v = v^{-1}u^{-1}a = (uv)^{-1}a = a\lambda_{uv}$; (ii) $\lambda_u = i$, if and only if $u^{-1}a = a$ for all $a \in G$, whence $u = 1$; (iii) $a\lambda_u\rho_x = u^{-1}ax = a\rho_x\lambda_u(a \in G)$; (iv) suppose that $a\theta\lambda_u = a\lambda_u\theta(\forall a, u \in G)$. Put $a = 1$ and define $x = 1\theta$. Then $x\lambda_u = 1\lambda_u\theta$, that is $u^{-1}x = u^{-1}\theta$. Since u^{-1} runs through G when u does, it follows that $\theta = \rho_x$. Similarly, when $a\eta\rho_x = a\rho_x\eta$, $\eta = \lambda_u$, where $1\eta = u^{-1}$.

(10) Put $[G:H] = n$. Then there exists an injective homomorphism $\theta: G \to S_n$. Hence $|G\theta| = 168 \leq n!$ and therefore $n \geq 6$.

(11) If the centre of the rectangle is at the origin, and the sides are parallel to the x-axis and y-axis respectively, then the symmetries are the identity and the rotations through π about any of the coordinate axes. The group is isomorphic with the four-group.

(12) In the coset expansion of G with respect to G_1, the letter 1 occurs precisely in all permutations that do not belong to G_1, that is $g - (g/n)$ times; the same is true for any other letter.

Chapter VIII

(1) The group V (p. 46) is a normal subgroup of A_4 of order 4 and is therefore the unique Sylow group of that order. Each 3-cycle generates a Sylow 3-group, for example 1, (123), (132). There are four such groups of order 3, each corresponding to a selection of three objects out of the four objects on which A_4 acts.

(2) The permutations $a = (1234)$ and $b = (24)$ generate a subgroup of order 8, comprising the permutations
(1), (1234), (1432), (24), (13), (12)(34), (13)(24), (14)(23).
This is a Sylow 2-group. Since $a^4 = b^2 = (ab)^2 = 1$, this group is isomorphic with the dihedral group (table xi, p. 51). The Sylow group is clearly not normal and hence not unique. There are three Sylow 2-groups.

(3) Such a group would have to possess eight groups of order 7 and seven groups of order 8, which is impossible in a group of order 56.

(4) There are $1 + xp$ Sylow p-groups and $1 + xp | p^2 q$. Hence $1 + xp | q$, which implies that $x = 0$. There are $1 + yp$ Sylow q-groups and $1 + yq | p^2 q$, $1 + yq | p^2$. Unless $y = 0$, this implies that $1 + yq$ is equal to p or to p^2; in either case $q | p^2 - 1$, which is excluded. Hence $G = P \times Q$, where $|P| = p^2$ and $|Q| = q$. Since P and Q are Abelian, so is G.

(5) Let $|G| = p^m g'$, where $(g', p) = 1$, and $|K| = p^\mu$. Use the double coset decomposition of G relative to K and any Sylow p-group P, say
$$G = Kt_1 P \cup Kt_2 P \cup \ldots \cup Kt_r P.$$
As in the proof of Theorem 29, it is shown that there exists at least one index j such that $|t_j^{-1} Pt_j \cap K| = p^\mu$, that is $K \leqq t_j^{-1} Pt_j$.

(6) By exercise (5), $t_j K t_j^{-1} \leqq P$. Since $K \lhd G$, $t_j K t_j^{-1} = K$, whence $K \leqq P$.

(7) Let $|G| = p^m s$, where $(p, s) = 1$. Then $|P| = p^m$. Now HP is a group because $H \lhd G$ and hence $HP = PH$ (Theorem 5, p. 52). Clearly $P \leqq HP$. Hence $|HP| = p^m t$, where $(p, t) = 1$, and $t | s$ by Lagrange's Theorem. The relation $HP/P \cong P/H \cap P$ (Theorem 10, p. 73) shows that HP/P is a p-group, since this is obvious for the right-hand side.
(i) It suffices to show that $|G/H| : |HP/H|$ is coprime with p; but this quotient is equal to $|G| : |HP| = s : t$, which is indeed coprime with p.
(ii) Again, by Theorem 10, $|H| : |H \cap P| = |HP| : |P| = t$, as required.

Bibliography

BURNSIDE, W., 1911. *Theory of groups of finite order*, 2nd edition. (Reprint by Dover Publications, 1955.)

COXETER, H. S. M., and MOSER, W. O., 1965. *Generators and relations for discrete groups*, 2nd edition (Springer).

HALL, MARSHALL JR., 1959. *The theory of groups* (Macmillan).

HUPERT, B., 1967. *Endliche Gruppen* I (Springer).

KUROSH, A. G., *The theory of groups*, 2 vols. (transl. from the Russian by K. A. Hirsch, Chelsea, 1955).

MILLER, G. A., BLICHFELD, H. F., and DICKSON, L. E., 1916. *Theory and application of finite groups* (John Wiley: reprint by Dover Publications, 1961).

ZASSENHAUS, H., *The theory of groups* (transl. from the German by S. Kraivety, 2nd edition New York, 1958).

Index